"*What about Evolution?* is one of those rare resources that sheds more light than heat when addressing the controversial issue of the relationship between evolution and faith. Writing from their own disciplinary areas of expertise, the authors guide us through a complex thicket of issues—biological, theological, biblical, and pastoral—with both wisdom and grace. An excellent resource for you—and surely others you know!"

—TODD WILSON, Cofounder & President,
The Center for Pastor Theologians

"A scientist, a theologian, and a pastor write about evolution. Instead of the entrée to a witty joke, *What about Evolution?* is a serious invitation to learn more about both science *and* the Bible. Humility, gracious dialogue, and fellowship take center stage as the authors respond to challenging questions from the science vs. faith conversation. This book supports anyone searching for faithful and fruitful ways to converse about evolution and Christianity."

—SARAH BODBYL ROELS, Faculty Developer,
Colorado School of Mines

"This book tackles some of the most difficult questions that emerge at the interface of evolution and Christianity. It is technically sound and nuanced in dealing with scientific, biblical, and theological issues, yet it remains accessible and pastoral at its core. The authors demonstrate the importance of engaging well with these issues and model how this can be done with humility in the context of a community with different experiences, perspectives, and concerns."

—RYAN BEBEJ, Associate Professor of Biology, Calvin University

What about Evolution?

What about Evolution?

A Biologist, Pastor, and Theologian Answer Your Questions

April Maskiewicz Cordero
Douglas Estes
Telford Work

CASCADE *Books* · Eugene, Oregon

WHAT ABOUT EVOLUTION?
A Biologist, Pastor, and Theologian Answer Your Questions

Cascade Books
An Imprint of Wipf and Stock Publishers
199 W. 8th Ave., Suite 3
Eugene, OR 97401

www.wipfandstock.com

PAPERBACK ISBN: 978-1-6667-1294-0
HARDCOVER ISBN: 978-1-6667-1295-7
EBOOK ISBN: 978-1-6667-1296-4

Cataloguing-in-Publication data:

Names: Cordero, April M., author. | Estes, Douglas, author. | Work, Telford, author.

Title: What about evolution? : a biologist, pastor, and theologian answer your questions / April M. Cordero, Douglas Estes, and Telford Work.

Description: Eugene, OR: Cascade Books, 2022 | Includes bibliographical references.

Identifiers: ISBN 978-1-6667-1294-0 (paperback) | ISBN 978-1-6667-1295-7 (hardcover) | ISBN 978-1-6667-1296-4 (ebook)

Subjects: LCSH: Evolution (Biology)—Religious aspects—Christianity. | Creation.

Classification: BS651 W4235 2022 (print) | BS651 (ebook)

Contributors

April M. Cordero is Professor of Biology and Dean of Educational Effectiveness at Point Loma Nazarene University. Her PhD from the University of California at San Diego focused on promoting learning in biology. April has been teaching since 1990, and currently teaches ecology and evolution to college science majors, nonmajors, and graduate students. She has earned three teaching awards, was a SCIO Visiting Scholar at Oxford University in 2015–2016, and is currently on the Board of Directors for BioLogos. She is also active in several professional development projects with schoolteachers as well as university biology faculty.

Douglas Estes is Associate Professor of Biblical Studies and Practical Theology at Tabor College. His PhD in theology from the University of Nottingham included emphases on biblical interpretation and science/religion. Douglas has written or edited ten books and published more than fifty essays, including *Braving the Future: Christian Faith in a World of Limitless Tech* (Herald, 2018) and fifteen science essays in *Christianity Today*. Prior to being a professor, he pastored for sixteen years, and he is a fellow at the Center for Pastor Theologians. Douglas's undergraduate degree was in chemistry, and he far preferred p-chem and instrumental analysis to organic chemistry.

Telford Work is Professor of Theology at Westmont College. His PhD in religion is from Duke University, focusing on doctrine of

Scripture. He teaches Christian doctrine and theology to under-graduates as well as interdisciplinary first-year seminars, and has taught on five continents as a visiting professor. His most recent book is *Jesus—The End and the Beginning*, and his course lectures on Christian doctrine and New Testament are online at YouTube and iTunes U. He is the Scoutmaster for his boys' Scout troop and on a team of volunteer chaplains for the Santa Barbara County Jail.

Table of Contents

Table of Contents

TABLE OF CONTENTS

Acknowledgments

April: A huge thank you to all of my friends, colleagues, and students who asked questions and dialogued with me about evolution and faith over the past fifteen years. It's been an honor and joy to engage with you and to learn from you. So many of you told me that you wanted or needed a short explanation of evolution and its compatibility with our faith, so I wrote this with you in mind. This book would not have come to be if it weren't for Dr. Jamie Jensen at BYU and all of her colleagues involved with the RecoEvo project. Telford and I met at the 2018 workshop, engaged in many stimulating conversations, and eventually motivated each other to create this book. Thank you Telford for taking care of all of the details and tasks with our publisher. Dr. Ryan Bebej and Dr. Sarah Bodbyl Roels, I am very grateful for the valuable comments, attention to detail, and the feedback you provided on drafts of my chapter. Almost all of the images throughout chapter 1 were created by Sandra Sketches, an excellent digital artist whom I've known since she was in high school. It was a pleasure working with you. A special note of gratitude goes to Dr. Dennis Venema for allowing me to use his language analogy and explanation, and to Dr. Gregg Davidson for allowing me to use his whale diagram. Finally, I would not have been able to get this work done without the support and encouragement of my sons, Jake and Tyler, my parents, Vince and Lynn, and my husband, Denton. Your support and love sustain me through everything I do.

Douglas: I would like to acknowledge and thank the many scientists and science-oriented people who have dialogued, discussed, and debated with me over the years all the wild and wonderful features of the world God created. Truly, it is a privilege to behold all that creation is. I thank all of those friends and acquaintances who work in science and technology for helping us to see and understand God's handiwork in increasingly meaningful ways. I especially single out Jason Woods, who made his career leading teams of scientists and technologists in the chemical industry. Jason and I grew up together, both majored in chemistry in college (at different colleges), and often had ferocious and wide-ranging debates about evolution, creation, God, and the sciences. [I made him listen to pre-*Jesus Freak* DC Talk Christian music, but he made me eat at Taco Bell, so we are even!] I would like to thank *Christianity Today*, Zondervan, Cascade, and Herald Press for accepting so many of my science and technology essays and books over the years—works that I hope might spur future generations of Christians to love and value science and tech as an outwork of God's plan for humanity. I also thank my mom, Nadine, who supported all of my endeavors even as she led NASA's Management Education Center for nearly thirty years. Most significantly, for all that I have accomplished I owe gratitude to my flawless wife, Noël, and my superlative, supportive children: Wyatt, Bridget, Violet, and Everett. In all that we do, *soli Deo gloria*!

Telford: I too am grateful for Jamie Jensen and the RecoEvo project, which invited teams of biologists, theologians, and pastors to discuss evolution and faith, and showed me just how powerful that combination is. Also for my Westmont colleague in biology Yi-Fan Lu, who invited me along with pastor Maurice Lee to that conference in 2018. And I'm particularly grateful to April and Douglas for taking a chance on turning that team structure into writing. Other past and present colleagues who have been invaluable in shaping my view of science-and-faith include Jeff Schloss, Nancey Murphy, Tom Fikes, Amos Yong, Jamie Smith, and Tremper Longman. Westmont College's professional development committee and provost's office supported my part in the project's writing and editing. At Cascade, all three of us writers appreciate Jesselyn Clapp's able and cheerful management of our unusual format and formatting, and Rodney Clapp's editorial eagle eye and support. Thanks, always, to my wife Kim and kids Jeremy, Daniel (and his new wife Megs!), Junia, and Ben. Finally, I'm so grateful for all—students, family, friends, pastors—who have wrestled with the challenge that biology and the gospel pose to each other, no matter what your positions are on either one. You've enriched and blessed me as we've partnered with one another in earnest inquiry into deep matters of organic life and eternal life. I look forward to more.

Introduction

WE WROTE THIS SHORT book for those who are either entering into the encounter of Christianity and evolutionary biology, or have someone they care about who is. Perhaps your scientifically minded, levelheaded child just joined a Christian tradition that affirms creationism, and you're scared that he or she is going to believe that the earth is only six thousand years old. Or perhaps you have always been curious about evolution, but learned from your church that you cannot be a Christian and accept evolution. Maybe your son or daughter whom you raised in a Christian tradition that rejects evolution is now learning about it from an experienced biologist, and you're worried what will happen to his or her faith. If the encounter is happening in college, we can imagine a Thanksgiving conversation that gets awkward:

Mom: And how are your classes going?

Student: I'm mainly in general ed courses. They're all right, I guess. My introduction to biology course is fascinating, though.

Uncle: Oh, how?

Student: Like how much human DNA comes from our ancestors. It's like 98 percent.

Uncle: Well, just because they are similar doesn't mean one species came from another.

Student: But you can trace mutations and see when changes occurred. Like when our ancestors lost our ability to make vitamin C about sixty million years ago. A mutation broke the gene sequence that used to make it. We still carry the broken version.

Uncle: (Silence.)

Mom: But that's just a theory, right?

Student (with irritation): Sure, but it explains a lot. Gravity is "just a theory" too.

We don't know how the rest of this conversation goes, but it doesn't sound easy for anyone. Maybe the uncle is quietly stewing about secular humanism invading his family and threatening a young person's faith. He could be rehearsing a rebuttal that appeals to God making the animals in Genesis 2 and fossils coming from Noah's flood. The student may be enjoying the rush of flexing some fresh young-adult power, irritated by boomer relatives' resistance, and yet inwardly saddened and frightened by the distance suddenly opening up among them. Mom may be resigned to the future she had dreaded and prayed over when saying goodbye at campus orientation over a year ago. Quietly skeptical, science-minded younger brother might be delighted to see an ally in the making, and watching closely to see what happens next.

My (Telford's) own background is more like the first scenario: I grew up in a casually liberal Protestant family where evolution was taken for granted. When I became a strong Christian and joined an evangelical church, it scared my family to death. The whole shift seemed to them beyond the pale, intellectually backward, even cultish. I absorbed the assumption that biblical faith was incompatible with human evolution, and I remember scaring one of them when I mentioned problems with evolution.

For me (April), I gave up my belief in God because I was told in college that evolution and faith were mutually exclusive. It was a difficult and lonely journey back to my faith because no one would talk to me about how to reconcile my evolution knowledge with my desire to follow Christ.

My situation (Douglas) was also different. I grew up in an environment where both of my parents worked at NASA; my father was not a believer but my mother was. I attended church somewhat regularly, and it was one that held to the typical caricatures of science and evolution. Still, for me science and religion have always peacefully coexisted. I saw the positives and negatives of both perspectives. Focusing on the sciences in high school and college gave me many opportunities to discuss and debate science and religion with both believers and nonbelievers.

None of our stories are simple. All involve negotiating evolution and faith in multiple, intersecting parts of our lives. We expect your situation is equally tangled.

Whatever situation you're facing, it's not right that anyone should have to walk through the evolution-faith journey alone. Shouldering the burden of representing material that's still new and introductory is difficult, and it's not right that you should face it unprepared. We wrote this little book to help prepare you for conversations about evolution and faith so that you can support everyone involved.

We three authors are excited to walk beside you as you think about evolution and faith. April is a biologist, Douglas is a pastor turned biblical scholar, and Telford is a theologian. We're all evangelical Christians of some kind. We assume that *someone* in your mix cares about Christian faith. We aren't assuming here that it's the student, you, a friend, or the biology professor. We *are* assuming that at least one of you is intrigued by how evolution and faith might harmonize or collide, and we're betting that someone perceives a conflict between evolution and Christian convictions.

I (Douglas) am passionate about helping Christians understand and appreciate the sciences. When studying the sciences, I worked with both believers and unbelievers. Sometimes the unbelievers would knock religion (and their nihilism was troubling), but everyone knew that I was a Christian and I was always clear about my spirituality, so I never experienced much personal criticism. When I switched to ministry in graduate school, I was discouraged to find that Christians would engage in personal criticism not just of science but also the people who worked in the sciences (even as their overall piety was encouraging). I realized that there was work needed for both groups.

I (April) chose to dive into the deep end of this evolution-and-faith tension about twelve years ago for a number of reasons, but mostly because of the stories my students share with me. My heart breaks when my college juniors and seniors tell me they no longer attend church, or worse, have abandoned their faith because they cannot resolve the tensions between their religious upbringing and science. Several recent studies confirm that many eighteen- to twenty-nine-year-old Americans are disengaging from their Christian faith communities, and one reason often cited is that they do not feel equipped to deal with the anti-scientific views of their churches. While college science majors receive instruction in biological issues that evoke controversy (such as evolution), these students are not usually well versed in the science-and-faith dialogue. Moreover, few parents, relatives, or friends are equipped to engage in substantive dialogue with these young adults to help them overcome the disconnect they are experiencing between what they hear at church or at home, and what they learn in science class. We hope this book helps you engage in those conversations well.

I (Telford) was drawn into this topic just by teaching at a Christian liberal arts college where I had colleagues with strong evangelical faith and deep commitments to both evolutionary biology and the authority of Scripture. Our college motto is from Colossians 1:17: "Christ holds preeminence in all things." So Jesus reigns in nature and the sciences that study it, in the humanities, in dorm and professional life, in the Scriptures of course—and discerning the signs of his reign is our responsibility as scholars of faith. Because we engage the Bible, Christian life, and the sciences together, that Thanksgiving conversation breaks out again and again in our classes, churches, and student families. Years of it have convinced me that evolution and classical Christian belief *do* interfere with one another, though much less than many people assume. Whether we human beings came from earlier primates or directly from dirt makes a difference. There are real stakes involved here, but they aren't necessarily where the caricatured positions see them.

We'll put our cards on the table and start with where the three of us *don't* see interference. We believe that if humans evolved from other species, there is still room for the God of Israel to be our Creator, for Jesus of Nazareth to be creation's risen Lord, Messiah, and Savior, for the Bible to be trustworthy and true, and for you and those you care about to be divinely loved and sealed for an eternal future in God's kingdom. In fact, there's more than just room; there's harmony.

We think there is as much room for classic Christian faith in an evolutionary scenario as there is for a Copernican one where the earth revolves around a sun that slowly circles from the Orion Arm of the Milky Way galaxy. I (Telford) got a long look at the Milky Way on a moonless night from eleven thousand feet while backpacking in the Sierras last summer. I wasn't imagining the kind of "firmament" or hard shell that dazzled my biblical ancestors when they gazed at the desert sky. I didn't imagine it all circling me as they did. No, to me that night sky appeared as an infinite expanse, whose faintly glowing "milky" mass was in fact 250 to five hundred million close neighbors obscuring the innumerable galaxies beyond—a humbling reality that I was almost always oblivious to thanks to the sun, the moon, and California air and light pollution. Here too there are stakes: How is it that the little spinning earth with its minor league sun on the fringes of just one of those galaxies can be the center stage of creation's ultimate drama? That question haunted the Hebrews too, as Psalm 8 demonstrates. And the church has successfully adjusted its imagination and found modern astronomy not to be the hindrance it once threatened to be. (Though in our Internet era, the numbers of flat-earthers are on the rise. That says more about how information travels nowadays than the state of contemporary astronomy.)

We three authors believe the situation with evolution is similar. The relationship between Darwinian evolutionary theory and classic Christianity does need attention, but it doesn't need to come down to an inevitable conflict. Furthermore, we think that's good news. Because while there are trade-offs involved, we don't think you or the person you love is facing a stark choice between embracing evolutionary biology and remaining in Christ's embrace.

However, in our current culture that may not be clear at all. Many evangelicals warn that embracing evolution amounts to a flat rejection of the Bible, of Jesus Christ's work as "last Adam" (see Romans 5:12–21), and of humanity's unique glory and place in creation—or at least puts us on a slippery slope that ends there. Many secular influencers say basically the same thing: not to warn believers, but to insist that God is dead and that Christianity has no rightful place in a truly scientific age. It didn't have to be this way. These voices emerged, especially in America, for particular historical reasons. Karl Giberson, a physicist and historian who is also a Christian, has written a compelling and helpful history of the conflict.[1] He shows how we got from the natural theology of Darwin's upbringing to an early twentieth century fundamentalism in which evolution was actually considered acceptable, then to the twentieth century's increasingly polarized divide between Christian creationists and intelligent design advocates and secular Darwinians. From inside, the conflict might seem inevitable and unavoidable, but in fact it is neither.

Here is how the three of us will proceed. Each of us will take up a series of questions—both common ones and less common but still important ones—and answer them from our own disciplinary standpoints. The other two authors will occasionally weigh in with observations and reactions in footnotes. We hope this gives you a better sense of how our different fields, vocations, and personalities shape our viewpoints, and what the give-and-take looks like among evangelicals on this topic. We won't always agree on particulars, and we make each other a little nervous once in a while, but that is true of Christian fellowship in other areas of life and learning, so why not this one too?

In chapter 1, April explains the basics of biological evolution using analogies and diagrams, referring to the data as much as possible. At an introductory level, she covers everything from "micro" to "macro" evolution, explains randomness, and discusses why evolution is taught in schools.

In chapter 2, Douglas takes up the general and spiritual concerns that many people have when they first start to encounter

1. Giberson, *Saving Darwin*.

both faith and evolution. He covers topics like how accepting evolution affects one's faith and how one can be a faithful Christian and someone who accepts evolution. His approach is often pragmatic, pointing out the value of learning about evolution even if one doesn't have all the answers to fully accept evolution.

In chapter 3, Telford fields questions at the intersection of theology and biology, especially at the pressure points where one poses a problem for the other. How do key Christian claims and teachings work if the universe really is the way evolutionary biologists say?

Christians look back romantically and wonder what it was like to live in pivotal moments in Christian history: "Bible times," of course, the evangelization of the Roman Empire, the eras of the Protestant Reformation and world missions, and the great persecutions that persist around the world to this day. If we brush away the romanticism, these times were difficult, treacherous, tense—and formative. Engaging in evolution and faith conversations can be the same. "May you live in interesting times" is said to be a Chinese curse. It's not *actually* a Chinese curse, but it would be an apt one. Peaceful times are much more pleasant. Yet living in interesting times is also a blessing. We've received that blessing whether we like it or not. The three of us hope our chapters help you face the challenge well and fruitfully.

A Biological Perspective

April Cordero

Why do we need to talk about evolution and faith?

I DON'T RECALL GIVING much thought to evolution while in high school, but during my first year in college at a large public university, I remember my biology professor telling the class that a person cannot believe in God and accept evolution. The basic argument was that evolution proved that there was no God. I found this puzzling, and I met with two different pastors who likewise told me that I must choose: either faith or evolution. The more I learned in biology courses, the more evidence pointed toward the reality of evolution. Since both sides were telling me that they were mutually exclusive, I made the conscious decision to give up my faith and any belief in a God or a higher power.

Fast forward to eighteen months after I graduated from college with a biology degree. It was 1989 and I, still an atheist, moved to Japan for a life adventure. My loneliness and boredom led me to read the only unread book written in English that I could find: the Bible. Reading from page one was a slow process, but thanks to my sister, who responded to many questions about the Old Testament via snail mail correspondence between Japan and the

United States, I made the decision to rededicate my life to being a Christ follower, and shortly thereafter I came back to the States. I spent the next decade hiding my love of all things biology from my church friends and hiding my Christian beliefs from my science colleagues because everyone was saying that evolution and faith were incompatible.

During the 1990s, learning how to live out the Christian command to love the Lord with my mind (Luke 10:27) presented the biggest challenge for me. I was successful at loving God with my heart, soul, and strength, but I had to find a way to reconcile my faith, the Bible, and my acceptance of evolution. At the time, none of the pastors I knew would open-mindedly discuss evolution with me, and I couldn't find any books on reconciling evolution and faith. Remember, this was before we could do a Google search.

Coming to a resolution was a slow and lonely process, but by the end of the 1990s, I came to understand that many parts of the Bible are interpreted today in ways that were not necessarily intended originally. For example, many biblical scholars agree that Genesis was not written as a scientific explanation of creation. This is supported by the fact that the orders of creation given in Genesis 1 and Genesis 2 are quite different. I also came to recognize that different books of the Bible were written at different times, for different groups of people, and that when viewed within their contexts, the messages from these books were even more beautiful and profound. I came to realize that taking the Bible on its own terms—not forcing it to answer questions or provide information it is not intended to address—allowed me to take science on its own terms. Two decades later, I consider it a privilege to speak at universities and churches about the compatibility of evolution and faith.

I've always heard that evolution proves there is no God. Is that true? Don't you have to choose between evolution or faith?

In churches and among Christian college students, it is still common to hear that evolution and faith are not compatible, that one must choose between them. But that is simply not true. There are millions of people—scientists, theologians, biblical scholars, and pastors included—who reconcile biological evolution with their faith. In fact, many of the most famous and influential Christian scholars and leaders of the last 150 years believed evolution was compatible with Christian faith. These include B. B. Warfield, C. S. Lewis, Karl Barth, Billy Graham, N. T. Wright, Pope Benedict XVI and Pope Francis, Philip Yancey, John Ortberg, and Tim Keller, to name just a few.[1] If we dig a bit deeper into the perception that evolution and faith contradict each other, we often find that people are making incorrect assumptions about both evolution and creation.

Biological evolution is completely neutral about a higher power. It neither supports nor denies a creator. In fact, science cannot prove or disprove the existence of God. When some people claim that evolution "proves" there is no creator, they are overstepping the boundaries of science. Science cannot answer such metaphysical questions as whether or not there is a God.[2]

A better way to think about evolution and creation is to focus on the fact that all Christians can agree that God created, and it was "good." In this way, we are all "creationists" because we believe in a divine power who created. What we may disagree on, however, is *how* God created—the mechanism of God's creation. Just because science can describe or explain how Earth came to be filled with diverse life-forms over four billion years does not leave any less room for God in the process. Science gives us the mechanism by which life develops, and Christianity gives us the agent behind that mechanism. When Christians speak about God as the creator, we are speaking about the entity behind the cause, not the causal mechanisms themselves.[3] Therefore, a Christian can

1. BioLogos.org, "Famous Christians Who Believed Evolution is Compatible with Christian Faith" and "B. B. Warfield, Biblical Inerrancy, and Evolution."

2. TW (for Telford Work; we will be commenting on each other's chapters): This is not because God is hiding or just a human fabrication, but because of the built-in boundaries of the scientific method. What makes science's tool kit so powerful in discovering within a range of kinds of knowledge also limits it beyond that range.

3. TW: I like to say that "creation" names some of the *relationships* we have

accept the science of evolution while still acknowledging God as the creator, because God is the agent and evolution is the process.

The theology is of course more complicated than my musings in these initial paragraphs, and this book's other two chapters dig much deeper into the compatibility of evolution and faith. Yet, one thing I've learned in my role as a biology professor trying to help hundreds of students reconcile evolution with their faith is that oftentimes students need to uncouple their personal theology about creation from their general theology about God, Jesus, and the Bible. As you read through this chapter, you might find this uncoupling helpful as well. As Christians, we can maintain our trust that God is the creator and savior while reexamining our beliefs about *how* God created. According to Saint Augustine (AD 354–430), our trust in the revelation of Scripture should not blind us to the revelation we can gain from the study of nature. Both areas of knowledge are worthy of full development. I firmly believe that following the path of truth, whatever its source, is a profound act of faith for the Christian.

What does the theory of biological evolution claim?

In its simplest form, the theory of biological evolution is a well-tested and well-confirmed explanation for the incredible diversity of species we see on Earth. The evidence for evolution reveals that the living organisms we see on Earth descended and diversified from earlier forms during the history of Earth. Evolution does not, however, answer "why" questions: *Why is there life on Earth? Why are we here? What is the purpose of life?* Evolution only describes the mechanisms that lead to the changes we see in living species over time.

To clarify this further, I need to describe what evolution does *not* address. Evolution is not an explanation for the origin of the

---

with the God of Israel, including some causal ones. Specific *ways* that God is responsible for us do matter, just as it matters whether a human parent is biological or adoptive, but affirming that God created everything doesn't lock us into affirming or denying specific mechanisms in that process.

universe, or even the origin of life on Earth. Questions about the big bang and the origin of life are important, but evolution as a scientific explanation begins once life originated, and explains how the earth has come to be filled with so many varied life-forms. To dig a bit deeper into the mechanisms of evolution, I'll begin with bacteria.

In September 2016, a video that showed the evolution of antibiotic resistant bacteria went viral on the Internet. The research was conducted in the Kishony Lab at Harvard Medical School. I highly recommend you pause and watch the three-minute video now.[4] The video begins by showing a giant petri dish divided into sections with each subsequent section having ten times the antibiotic dosage as the previous. On day one, none of the bacteria are resistant to the antibiotic in the first section, but after a few days as existing bacteria divide, minor genetic modifications occur in some of the bacteria "offspring" that allow them to survive in the higher dose. These resistant bacteria divide, passing on their new genetic material, and fill up the next section of the dish. As more time passes and bacteria continue replicating, again some new bacteria with minor genetic modifications allow them to resist an even higher level of antibiotic. This new generation moves into the next section of the dish and begins replicating, passing on the resistance gene and again populating that section. After eleven days, all the sections of the dish are filled with bacteria including the section with one thousand times the initial dosage of antibiotic. In short, the researchers showed that as bacteria encountered ever higher doses of antibiotics, new genetic material emerged allowing new generations of bacteria to resist higher and higher concentrations of antibiotic in only eleven days. In simple terms, the bacteria population evolved.

A very important point to understand is that no individual bacterium changed. It is the bacterial "offspring" that have a minor genetic difference from the parent bacteria. So the changes that we see as a result of evolution are not within individuals, but occur

4. Harvard Medical School, "Evolution of Bacteria on a 'Mega-Plate' Petri Dish (Kishony Lab)." TW: It's definitely worth your time!

as a population changes across generations. New generations may be slightly different, genetically, from previous generations. If you were to compare any one generation of bacteria to the next, the individual bacteria would seem almost identical to the previous generation. But over many, many, many generations, thousands of genetic differences accumulate. While small incremental changes are barely noticeable from one generation to the next, if given enough generations, the change that occurs in the population can be quite dramatic.

Some would say that this kind of evolution is not controversial at all because a species is merely changing in a minor way. In fact, almost all, if not all Christians accept this level of evolution. We all acknowledge "superbugs" in hospitals that are resistant to all of the typical antibiotics. And most of us get a flu shot every year because we understand that flu viruses mutate. The real controversy arises when we talk about large-scale evolution, or what many call macroevolution.

What's the difference between micro- and macroevolution? Can I affirm some parts of evolution without affirming all parts?

The terms *microevolution* and *macroevolution* are often used by nonscientists to represent something much broader than their scientific meaning. I will use the terms *small-scale* and *large-scale evolution* instead, because they better represent the science. The bacteria example provided earlier would be considered small-scale evolution because, over a short period of time, we see changes in the frequency of particular genes in a bacterial population. Large-scale evolution, on the other hand, represents the idea that over very long periods of time, hundreds of millions of years, we see many unique species emerge and diversify. Large-scale evolution explains major trends in the diversity of organisms such as the proliferation of flowering plants and the transition from water animals to land animals.

According to the theory of biological evolution, the variety of life we see on Earth is the result of evolution from a common ancestor. Some Christians reject the idea of common ancestry while accepting that evolution can happen on a small scale (microevolution), as with bacteria. The problem with accepting small-scale evolution and rejecting large-scale evolution (macroevolution) is that the mechanisms that result in antibiotic-resistant bacteria are exactly the same mechanisms that lead to the large-scale changes we see over very long periods of time, like the origin of mammals. There is no difference in the processes except the amount of time considered. If given enough time, small genetic changes accumulate, resulting in the incredible diversity we see on Earth. For this reason, the terms *small-* and *large-scale evolution* better represent the idea that the mechanisms for evolution are the same in the short term as in the long term.

Related to this question is the perception by many Christians that scientists are divided about whether all life shares a common ancestor. They are not. Ninety-eight percent of professional scientists agree that all life on Earth, including humans, shares a common ancestor.[5] This extremely large majority of scientists accept large-scale evolution because the evidence is abundant, consistent, and credible. According to the National Academy of Sciences, "evolution itself has been so thoroughly tested that biologists are no longer examining whether evolution has occurred and is continuing to occur. Similarly, biologists no longer debate many of the mechanisms responsible for evolution."[6] While 98 percent of scientists have come to consensus about evolution and common ancestry, a Pew research study in 2014 showed that 46 percent of frequent churchgoers think that there is no consensus among scientists about human evolution.[7] My own informal survey in December 2018 at an Urbana conference with an audience of over two hundred eighteen- to twenty-five-year-olds revealed that 49 percent believed scientists were divided on the issue of human

5. Pew Research Center, "Major Gaps."
6. National Academy of Sciences, "Frequently Asked Questions."
7. Pew Research Center, "Strong Role of Religion."

evolution. This finding is disconcerting because there *is* consensus among scientists on human evolution.[8]

This brings us back to the original question. I surmise that because many Christians think scientists are divided about evolution, then it feels okay to affirm some parts of evolution but not others (to accept microevolution while rejecting macroevolution). But scientists are not divided. With no real difference between small- and large-scale evolution except time, it's not possible to justify accepting one and not the other on scientific grounds alone.

We've never seen cats turn into dogs (or apes into humans, or reptiles into birds), so why do scientists think organisms evolve into new kinds of organisms?

I have been asked some version of this question hundreds of times by university students as well as audience members when I speak at churches. The challenge in answering is that the question requires an explanation of two different things. First, we don't see large-scale changes in natural populations of *animals* over short periods of time because changes accumulate gradually. Thousands to millions of generations usually have to pass for the accumulation of mutations to lead to distinct differences that we might visually notice. Second, we certainly would never see cats turn into dogs (or frogs turn into snakes, or whatever animal species are compared) even over long periods of time, because this is not reflecting what animal evolution evidence shows us. I will tackle each of these explanations separately.

Evolution results from gradual genetic changes in a population over many, many generations. In the bacteria example, we were able to see changes in only eleven days because the generation time for bacteria is short; most bacteria can double every five to twenty minutes. In eleven days, somewhere between eight hundred to 3,200 generations passed. That's a lot of generations to accumulate genetic differences. For animals, however, generation

8. Pew Research Center, "Scientific Achievements."

times aren't minutes, but days and months to years or decades. When biologists say "gradual changes" over time, they are usually referring to extremely long periods of time, millions to hundreds of millions of years. For humans, a generation is about twenty-five years, so 3,200 generations would be about 80,000 years.

Dennis Venema, a geneticist and biology professor, uses an analogy about changes in the English language that can be very helpful when explaining the substantial changes we see within populations.[9] Take a look at the table below, where he compares the written versions of John 1:29 in Bibles over the past thousand years. While we can barely read the first version, the most recent (2011) is simple to understand.

	John 1:29
West Saxon Gospels, c. 990:	Oþre dæge Iohannes geseah þone Hælend to hym cumende, and cwæð: her ys Godes Lāmb; her ys se þe dēð aweg middan-eardes synne.
(transliterated)	Othre dage Iohannes geseah thone Heelend to hym cumende, and quath: her ys Godes Lamb; her ys se the deeth aweg middan-eirdes synne.
1395 Wycliffe Bible:	Anothir day Joon say Jhesu comynge to hym, and he seide, Lo! the lomb of God; lo! he that doith awei the synnes of the world.
1525 Tyndale New Testament:	The nexte daye Iohn sawe Iesus commyge vnto him and sayde: beholde the lambe of God which taketh awaye the synne of the worlde.
1611 King James Version:	The next day Iohn seeth Iesus coming vnto him, and saith, Behold the Lambe of God, which taketh away the sinne of the world.
2011 New International Version:	The next day John saw Jesus coming toward him and said, "Look, the Lamb of God, who takes away the sin of the world!"

Venema also shows some changes in individual words over time:

9. Venema, "How Language Evolution Helps Us Understand."

9

cwæð (pronounced like quath) → seide → sayde → saith → said

synne → synnes → synne → sinne → sin

to hym cumende → comynge to hym → commyge vnto him →
coming unto him → coming toward him

While each word has only minor changes at each transition, these gradual word changes contribute to the overall transformation of the sentence in the long term. And if we compare the verse from the tenth century to the twenty-first century, the difference is dramatic, yet the changes were gradual:

Oþre dæge Iohannes geseah þone Hælend to hym cumende, and cwæð:
her ys Godes Lāmb; her ys se þe dēð aweg middan-eardes synne.

↓ one thousand years

The next day John saw Jesus coming toward him and said, "Look, the
Lamb of God, who takes away the sin of the world!"

To connect this analogy to biology,[10] consider a population of squirrels that gets separated by the emergence of a steep canyon after a massive flood. Each separate population of squirrels would continue to reproduce and accumulate small genetic changes over multiple generations, but the genetic mutations that accumulate would be different for each population (because mutations are random—I'll explain that in another section). Eventually, after enough generations have passed, a comparison of the genes for these two populations could reveal that they diverged enough to become two species.[11]

10. DE: A word of caution about the limitations of analogies from two different fields of study: often these analogies work well on a surface level but fall apart on close scrutiny. A great example of this are the many analogies that Christians use from the natural world to explain the Trinity. In my opinion, this analogy works well on a surface level but would not survive close scrutiny.

11. TW: This happens all the time with island populations. The Channel Islands off the California coast host a number of dwarf species that evolved from their mainland ancestors, often because they were protected from mainland predators. Australia, Hawaii, and of course the Galapagos offer many more dramatic examples. The phenomenon is worldwide. It poses a problem

When I say "two species," I am not always referring to something as distinct as cats and dogs. A biological species is defined as a group of organisms that can reproduce with one another and produce fertile offspring. If the organisms in one population are incapable of reproducing with organisms in another population, they would be considered two distinct species. In the squirrel example, when the two populations are separated by a river for thousands of generations, the two squirrel populations could have accumulated enough distinct genetic differences that if a bridge were built across the canyon, the squirrels from each side would not be able to mate with each other. Biologists would then consider these two squirrel populations two different species. And given enough generations apart, the two populations could eventually begin to accumulate unique physical changes as a result of their different genetic mutations (longer legs, shorter tail, lengthened ears, a slightly modified mating scent, etc.). The passing of hundreds or thousands of generations could result in two visually distinct species.

So how does this relate to the dramatic differences we see between animals like cats and dogs? To explain that, we need to lengthen the time to millions of generations. Cats and dogs are related to one another by a common ancestor from the distant past. They share this ancestry with a whole suite of other animals, large and small, who all belong to the order Carnivora (see figure 1 below). Over 43 million years of reproduction and passing on genes led to a divergence of species including today's bears, hyenas, mongooses, skunks and more, as well as some marine mammals including seals, walruses, and sea lions. One early branch of the Carnivora led to two well-defined groups, the Caniformia (with canines included in this group), and the Feliformia (felines are included in this group). As you can see from the figure, cats (represented by the panther) do not give rise to dogs (represented by the coyote) or vice versa. Instead, they both share a common

---

for literalistic interpreters of Genesis 1–11 (How did Noah collect all those pairs of species in the first place? How did they make it all the way back to their habitats, and only those habitats?), whereas it strengthens the case for evolution.

ancestor that is neither a cat nor a dog, but is a carnivorous mammal in the distant past.

FIGURE 1

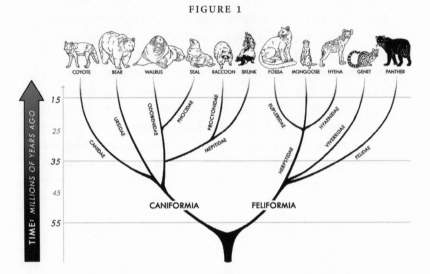

In summary, (1) evolution results from changes in the genetic material in a population over many, many generations; (2) individuals that are one or two generations apart may not appear to be different, but over many, many generations, their differences accumulate; and (3) large-scale changes that we see among species took millions of years.

Isn't evolution random?

In biology, when we use the term *random*, we mean something is extremely unpredictable. Unfortunately, evolution is often mistakenly held to complete randomness. I like how biologist Kathryn Applegate talks about randomness: "Far from being an indication of a 'godless' universe, one might conclude that randomness is one of God's favorite mechanisms for creating and sustaining life!"[12]

12. For an excellent discussion of randomness, see Applegate, "Understanding Randomness."

While some mechanisms of evolution *are* random, or occur by chance, other mechanisms including natural selection are not. Consider this example: Imagine a peacock born with a particularly large plume of tail feathers that causes the peacock to run slower than its muster (the word for a group of peacocks). A predator would likely be able to catch and kill this peacock before any others. In this case, if the peacock dies before mating, we would say that an extra large feather plume was *selected against* because those genes did not get passed on to the next generation. On the other hand, the peacocks that survive long enough to reproduce and pass on their genes to the next generation are *selected for*.[13] Natural selection, therefore, is not random; it is possible to predict which organisms might be better suited to survive and reproduce when we consider the environment they live in.

Genetic mutations, on the other hand, *are* random. When discussing evolution, "mutation" refers to changes in the genetic code—the DNA. DNA is comprised of four molecules referred to as nucleotide bases and represented by letters: A = adenine, C = cytosine, G = guanine, and T = thymine. Inside cells, these nucleotide bases pair up, creating the iconic double-stranded DNA (see figure 2 below). The human genome is made of 3.2 billion bases of DNA. Corn has 2.5 billion bases, but the number of bases is not as important as the order. The order or sequence of the nucleotide bases forms the instructions in the genome. Differences in the sequences are the main cause of diversity of life on earth.

13. TW: Philosopher Nancey Murphy complains that the phrase "natural selection" is a little misleading: selection implies a purposeful choosing agent, whereas "nature" is impersonal, thus neither an agent nor purposeful. One might as well say that rivers select lakes and seas. In these and other ways, biology uses terms metaphorically that suggest a conflict between nature's purposes and God's purposes. The Bible and theology use terms metaphorically too—we all do—in ways that can confuse us. See Murphy, "Science and Society."

FIGURE 2

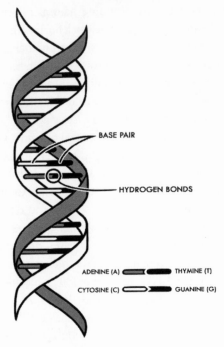

When cells replicate, there can be errors (or mutations) as the DNA is copied. Below is an example of a mutation. Each letter below represents one of the four nucleotide base molecules.

ACCGTTAGGA → AC**G**GTTAGGA

Notice that the third base changes from a "C" to a "G," cytosine to a guanine. This is called a substitution. The three types of mutations include *substitutions* (where one or a set of bases is replaced by different bases) as in the example above, *deletions* (a mistake in the replication of DNA that results in one or more bases getting deleted), and *insertions* (the addition of bases into the genome). These mutations occur during DNA replication, division, and re-combination, but can also be induced by environmental factors such as sunlight and other types of radiation. While mutations, or copying errors, are random with respect to whether or not they benefit the organism, not all types of mutations occur with equal probability. Genetic research shows us that some mutations are more likely than others, depending on the location in the genome.

Mutations can be helpful to an individual (e.g., for survival or finding mates) while others may be harmful, but most mutations are silent, meaning they have no effect on the next generation's structure (anatomy or physiology). Yet the differences we see among generations over time and the diversity of life we see on earth are the results of modifications to the DNA sequences that get passed on (for example, in sperm and egg).[14] Both helpful and harmful mutations can be acted upon by natural selection. In other words, they can be selected for or against by processes in nature (competition, predation, nonrandom mating, etc.). In short, mutations in DNA sequences occur at random when DNA replicates, and these mutations can result in anatomical or physiological differences that natural selection acts upon.

Two other mechanisms of evolution, migration and mating preference, are not random, while genetic drift *is* a random mechanism that results in modifications to populations over time, or evolution. Instead of explaining each of these additional mechanisms in detail, I've come to realize that for most Christians, the resistance to randomness has much more to do with the notion of purposeless change than a desire to understand the biology. While the other chapters will discuss randomness from the perspectives of a theologian and a pastor, it's worth noting here that evolution is viewed by some scientists as constrained by Earth's environment, and as such, can be considered directional and not purposeless. Simon Conway Morris, a Christian professor and chair of evolutionary paleobiology at Cambridge, provides an elegant, purposeful explanation of convergence as a mechanism of evolution that aligns well with the view that God's act of creation *is* intentional. In his introduction to his book *Life's Solution*, Morris writes: "What was impossible billions of years ago becomes increasingly inevitable: evolution has trajectories (trends, if you prefer) and progress

14. Viruses and some single-celled organisms have RNA instead of DNA, and RNA can also mutate.

is not some noxious by-product of the terminally optimistic, but simply part of our reality."[15]

Is there any evidence for large-scale evolution? And how can we know when nobody was around to document evidence for these kinds of changes?

The supporting evidence for large-scale evolution is abundant, robust, and corroborated by a variety of different fields of science including, but not limited to, anatomy, biogeography, molecular biology, paleontology, biochemistry, embryology, geochemistry, geology, and biological anthropology. There are hundreds of thousands of studies that confirm that all life on Earth shares a common ancestor. Here I describe just one well-documented case study, on the evolution of whales, to show you the depth of research and explain the supporting evidence from independent scientific fields.

As you may recall from your high school biology course, fish are not mammals, but whales and dolphins are. They are warm-blooded, breathe air through blow holes rather than gills, and give birth to live young and nurse them. Evidence shows that whales evolved from four-legged mammals beginning over fifty million years ago. In short, some land mammals, over long periods of time, moved back into the water. While this may sound ridiculous to you, scientists from independent fields of study provide an abundance of corroborating evidence for whale evolution. With limited space here, I will only briefly describe the evidence from paleontology (fossils), embryology (the study of embryos and their development), geochemistry (isotope ratios), and genetics.

Thousands of fossil specimens have been discovered of intermediate and transitional species that link four-legged mammals that lived on land to modern-day whales. The image below, drawn by Gregg Davidson, shows some of the fossil evidence for

15. TW: DNA's combination of continuity and mutability do seem well suited to the earth's changing and wildly diverse habitats in a way that theology labels "providential": a manifestation of God's wise provision for his creation. Life-forms that couldn't adapt through reproduction and mutation would die out with nothing to replace them.

intermediate species that indicate a series of adaptations from more terrestrial to more aquatic environments. Keep in mind as you examine these intermediate species what you learned earlier from bacterial evolution: any individual would look very similar to its parent and offspring, but over very long periods of time the gradual modifications in DNA accumulate. A collapsed diagram like this makes it deceptively look like major jumps in physical form.

FIGURE 3

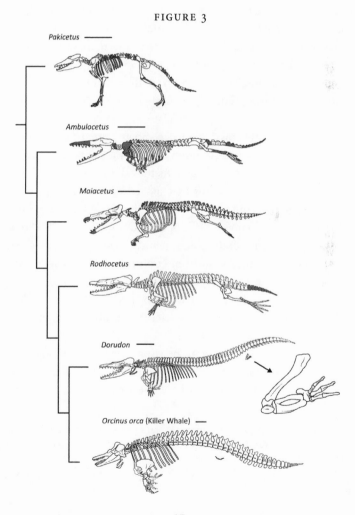

Pakicetus

Ambulocetus

Maiacetus

Rodhocetus

Dorudon

Orcinus orca (Killer Whale)

This diagram shows representatives of different species in the ancient ancestry of modern-day whales and dolphins, but it does not show the many other branches of whale relatives that have gone extinct. You will notice in the diagram gaps in the whale fossil record because not all intermediate species are known. Fossilization is a rare event because it requires just the right environmental conditions in order to take place. Nevertheless, there are a substantial number of fossils to support that whales evolved from land mammals, and no reliable fossils have been found to refute this claim. For example, we have never found any large whalelike aquatic mammal fossils before fifty million years ago. Based on the fossils we *have* found, paleontologists can hypothesize what the shape and size of various intermediate species might look like, and when additional fossils are discovered, they share much in common with what was predicted, including intermediate anatomical structures.[16]

This cumulative fossil evidence shows some of the physical changes we see in whale evolution that led to adaptation to life in water, including spine modifications that allow for more efficient modes of swimming; nostrils that moved toward the top of the skull and later became a blow hole; and the shrinking of hind legs to where they could no longer support the animal on land.[17] Fossil teeth provide corroborating geochemical evidence (e.g., isotope ratios) showing a transition from consuming fresh water to salt water during the earliest stages of this transition.

While the fossil data is fun to look at because it's visually clear, evidence from other domains of science definitively support that whales evolved from land mammals. Embryology evidence reveals that most whales, while still in the womb, begin to develop body hair but then lose it and are born without hair except for a small amount at their snout and blowhole. We see hind limb buds form

---

16. TW: This is important: it means evolution has predictive power as well as explanatory power. These are hallmarks of a successful scientific theory, as well as what the Bible calls wisdom.

17. TW: Who knew that Darwin-fish-with-feet on the bumper sticker is really a whale?

for a while in utero, but then disappear later during development. The whale embryo starts off with its nostrils in the usual place for mammals, at the tip of the snout, but during development, we can observe the nostrils migrating to their final place at the top of the head. This evidence confirms that whales possess the genes for features that connect them with terrestrial mammalian ancestors.

You might not view this data as irrefutable evidence for common ancestry or large-scale evolution. One could argue that God could have designed whales to have all of these similar traits as other animals because why create something new?[18] However, more recent findings looking at the DNA in modern-day whales reveal adaptations for terrestrial life that even more clearly point toward common ancestry.

For most mammals, olfaction (the sense of smell) is one of the most important senses, and odors in the air are detected by different receptors in the nasal cavity. Animals depend on a variety of these receptors to pick up different scents and process what they mean: danger, food, and communication with others. We find that much of the anatomical hardware needed for olfaction is missing in toothed whales (e.g., dolphins, porpoises, sperm whales, and killer whales). In other words, they do not develop the nerves and receptor cells required to detect and process smells in the air. That makes sense, because they live in the water. When we look at whale DNA, however, we see the genetic markers for air-based olfaction, and this can only be explained by common descent. In short, whales have the DNA to make olfactory receptors, but the DNA is defective. The genetic sequences exist, but are slightly modified such that the olfactory receptors do not develop. By comparing DNA sequences for olfaction in a variety of mammals alive today we see some specific genetic modifications in the whale DNA only.

18. TW: This is a popular response among creationists, but what are those vestigial hind leg bones doing deep inside the Dorudon and Orca?! This kind of reasoning is utterly unconvincing to people who don't already share creationists' assumptions about how to understand Genesis 1–2. Many of those people reject the whole Christian message as a result.

FIGURE 4

```
        Mouse: AAAAAAATGGATCCACTGCTTTGAAGGAGTCACCTGCATTATATTTTGC
          Cow: AAAGAAATGGATTCACTGCTTTGAAGGAGTTACATGCATTATATTTTGT
  Hippopotamus: AAAGAAATGGATTCACTGCTTTGAAGGAGTTACATGCATTATATTTTGT
   Minke whale: AAAGAAATGGATTCACTAC-TTGAAGGAGTTACATGAATTATATTTTGT
```

Look, a deletion in the gene!   *And a substitution!*

What we see in today's whales are nonfunctional genes that almost perfectly align with genes in other animals that *are* functional. The best explanation for why 65 to 80 percent of the olfactory genes in whales are defective is that as whales evolved from land mammals over time, detrimental mutations that made olfactory genes defective did not affect whale survival in the water nor reproductive success. In evolutionary terms we would say that the defective genes were not *selected against*, and consequently were passed on to the next generation. (Interestingly, sea lions and sea turtles have maintained functional olfactory receptor genes likely because of their reliance on land habitat for breeding and other important roles.) This DNA evidence strongly supports all of the other independent lines of evidence confirming that modern-day whales share a common ancestor with land mammals and evolved from them.

This one case study is only an ice cube on the tip of an iceberg of evidence for large-scale evolution. There are hundreds of similar case studies including mammal evolution, animals' transition from water to land, the divergence of plants, horse evolution, and the evolution of birds, to name just a few. The power of this evidence lies in the mutually reinforcing independent fields of study coming to the same conclusions.[19]

19. For additional evidence of various case studies of large-scale evolution, see *Scientific American*'s "Evolution" series online, as well as Shubin, *Your Inner Fish* and Asher, *Evolution and Belief.*

Doesn't evolution say we evolved from apes?

Humans did not evolve from apes, or gorillas, or chimpanzees, or any living primate you can name. That iconic image of a chimp turning into a gorilla turning into a human does not represent scientific evidence. Evolution does suggest, however, that we share a common ancestor with these primates. Just like you share a common ancestor with your second cousin once removed, this does not mean that you "came from" your second cousin. Instead, somewhere further back in your family tree, you share an older common relative. This analogy isn't completely accurate because you and your relatives are *Homo sapiens*; you are all the same species. When we talk about common ancestry for life on Earth, we are referring to relatedness among *different* species.

The evolutionary tree below shows a historical pattern of ancestry, divergence, and descent among primates.

FIGURE 5

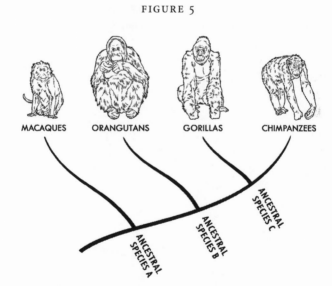

Each primate depicted at the top of the tree is a different species, meaning that they cannot mate and produce viable, fertile offspring with each other. We see that macaques, orangutans, gorillas, and chimps all share a single common ancestor at the deepest node, also known as the "root" of the tree, labeled here as Ancestral Species A. Species A is no longer alive today, but approximately fifteen million years ago it gave rise to generations of primates that genetically diverged. The ancestors of Species A alive today form four different groups as depicted in this tree. Similarly, orangutans, chimps, and gorillas share a more recent common ancestor, Species B, who is no longer alive today. The subsequent generations of population B branched off, resulting in three distinct species. Finally, gorillas and chimps share the most recent common ancestor, Species C, that is no longer alive, but gave rise to individuals that diverged to become today's gorillas and chimps. Each branch of this tree reflects the independent evolution of the lineages that have occurred after their divergence from a given common ancestor.

Scientific evidence also shows that humans evolved in this same tree cluster with other primates. In the diagram below, we see that chimps and humans share the most recent common ancestor, Species D, and thus are most closely related, genetically.

FIGURE 6

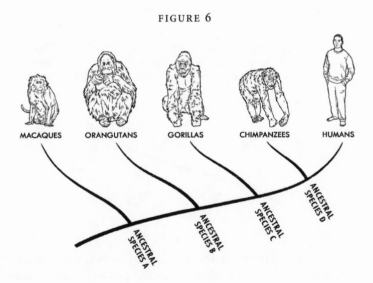

At this point, many Christians cringe and resist the idea that humans evolved in the same way as other species on Earth. Some Christians believe that if we evolved, this would make us less special because we are not uniquely created "from the dust." While this is a topic for theologians and pastors, and both of my coauthors will address this concern in their chapters, I offer here that it's not *how* we were created that leads us to believe we are special. We are special because we are loved by God and we can respond in relationship to God.[20] As Christians, we also believe that Jesus came as a human being and died on the cross for us, and these are the things that make us special. Accepting biological evolution as the explanation for our existence on Earth does not conflict with that belief. I'll leave the rest of this discussion to the specialists in the other two chapters. Here, I present some of the robust evidence that supports human evolution.

Just as with whales, an abundance of corroborating evidence from a variety of fields of science supports the evolution of hominins which include chimps, bonobos, and humans as well as all of the fossil descendants that came from the hominin common ancestor, including Australopithecus, Neanderthals, etc. Sometimes when people think about human evolution, they think of it as a linear path where one hominin group gives rise to the next one like rungs on a ladder. But this is not accurate. The evolutionary lineage for *Homo sapiens* is not like a ladder, but more like a tree or bush. As you see in Figure 7 below, many hominins lived on earth for long periods of time and many of these hominins lived during the *same* time periods. And just as we saw with antibiotic-resistant bacteria and the whales, we don't see a moment where a dramatic change or mutation led to the *Homo sapiens*. Instead, the process was gradual, and it took hundreds and thousands of generations to see a distinct difference between hominin forms.

20. TW: I'd put it only slightly differently: however we got here, we've been invited into a relationship of uniquely imaging the triune God as his heirs, which is fulfilled when we inherit the only Son's glory as co-heirs of his kingdom. I don't know whether other biological species are *capable* of some sort of relationship involving some sort of knowledge, or even enjoy one, let alone how they relate to the kingdom's coming new creation.

FIGURE 7

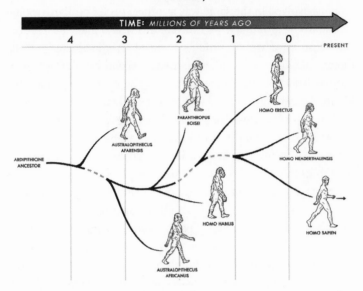

In the past twenty-five years it's been exciting to see a dramatic increase in the fossil evidence for hominin evolution.[21] Paleontologists and anthropologists have found literally thousands of diverse skulls and fragments of skeletons showing intermediate and transitional forms. I recommend spending some time exploring the Smithsonian website to see this vast set of fossils (search "Smithsonian human evolution evidence"). The fossil data reveal that over the last several million years, hominins gradually transitioned toward bipedal walking, greater height, and greater brain size. While there are many intermediate species that we call hominins, the precise relatedness between them is still being worked out as new hominin fossils and entire new hominin species continue to be discovered.

Just as with whales, the fossil data tell a clear story, but the best common ancestry evidence for human evolution comes from

21. TW: What if newer fossil evidence, or DNA evidence that was unknown in Darwin's day, had conflicted with common ancestry, or pointed in a number of contradictory or incompatible directions? That would have vindicated evolution's critics and called for some new theory. Instead, new information broadly confirmed it and explained some of its mechanisms.

more recent studies in genetics. Since I previously used olfactory genes as an example for whale evolution, I use them again here to represent one category of the thousands of genes that support common ancestry among humans and other animals. In most mammals like dogs and mice, more than 80 percent of their olfactory genes are functional, meaning that these genes provide the instructions to develop olfactory receptors that lead to the ability to sense all sorts of smells. In humans, almost all of the odor-receptor genes that we see in other mammals are still present in our genome, but 60 percent (about 600) of those genes are defective; they no longer provide the accurate instructions to make olfactory receptors (mutations have occurred). For humans, olfaction (or sense of smell) is not our most important sense. Instead, we rely primarily on sight and only secondarily on our other senses. This means that minor mutations modified the olfactory genes over millions of years, but these modifications were not selected against because they did not harm survival. We see the remnants of these defective genes in our genome today.

If we examine the family tree for primates, a clear pattern of "broken" or defective genes supports the claim that humans and other primates share a common ancestor. Figure 8 represents research on olfactory (smell) receptor genes among hominids.[22] Roman numerals I–V represent a mutation that occurred on a particular gene in an ancestral species. For example, in gene I, a mutation occurred before Ancestral Species A, the common ancestor to all five species, therefore, any species emerging from this common ancestor would have the mutation. In this case, all five species have this mutation, represented by the word *defective* in the table. In gene II, a mutation occurred after Ancestral Species A existed, and therefore only affected the ancestors of orangutans, gorillas, chimps, and humans, and not macaques. Similarly, the mutation in gene III occurred more recently, after Ancestral Species B, and affects only gorillas, chimps, and humans. And so on.

22. Gilad et al., "Human Specific Loss of Olfactory Receptor Genes."

FIGURE 8

GENE	MACAQUE	ORANGUTAN	GORILLA	CHIMPANZEE	HUMAN
V	FUNCTIONAL	FUNCTIONAL	FUNCTIONAL	FUNCTIONAL	DEFECTIVE
IV	FUNCTIONAL	FUNCTIONAL	FUNCTIONAL	DEFECTIVE	DEFECTIVE
III	FUNCTIONAL	FUNCTIONAL	DEFECTIVE	DEFECTIVE	DEFECTIVE
II	FUNCTIONAL	DEFECTIVE	DEFECTIVE	DEFECTIVE	DEFECTIVE
I	DEFECTIVE	DEFECTIVE	DEFECTIVE	DEFECTIVE	DEFECTIVE

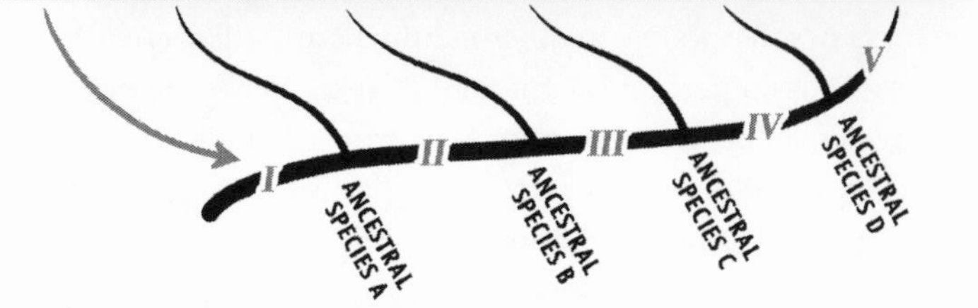

Why is this evidence for common ancestry? Consider mutation IV. We only see it in chimps and humans; no other primate here has this mutation. The simplest way to explain this is that it occurred in a common ancestor that gave rise to both species. Defective genes passed along in this same pattern are commonplace and provide solid evidence for common ancestry among other primates and humans. An abundance of independent lines of evidence points to the same conclusion, that humans evolved in the primate lineage.

So what makes us distinctly human? It's obvious that we are unique when we compare our accomplishments to the rest of life on earth. But what exactly is that distinctiveness? Some scientists argue that one distinctly unique feature of *Homo sapiens* is our theory of mind (ToM). ToM involves the ability to attribute mental states to oneself and others (e.g., understanding others' actions, feelings, and intentions). In other words, I can think about what you must be thinking about this topic. There are numerous other related claims for human distinctiveness, including our ability to perform highly cognitive tasks and behaviors such as language and abstract thinking, our cultural diversity, our empathy and morality, our use of symbols, our ability to think about our present, past, and future, and our ability to cooperate with each other, just to name a few. As a Christian, this research might intersect with how you think about being made "in the image of God."[23]

Isn't evolution just a theory?

Let's look into the meaning of *theory*. In our everyday language, we use the term *theory* to refer to a guess or conjecture. For example, "I have a theory for why my son continuously exceeds his curfew by ten minutes," or "I have a theory for why so many people dislike pop music." Popular usage of theory treats it as a proposition without much more support than a hunch. In science, however, theory means something *very* different. A scientific theory is an explanation of some aspect of the natural world that has been substantiated through repeated experiments or testing. A theory in science provides a framework that ties together evidence substantiated by many, many researchers' observations and peer-reviewed publications.

Sometimes people confuse a scientific theory with a hypothesis. A hypothesis is a statement or explanation that is speculative; for example, "Soil that includes compost made from kitchen scraps improves plant growth." Experiments and observations are then used to confirm or disconfirm the hypothesis. In this example, you would mix your countertop kitchen scrap compost into your plant soil when you plant new herbs. Then you would compare the growth of those herbs with herbs grown without compost added to the soil. If the compost soil plants are larger or healthier, you might conclude that your hypothesis is supported. But this does not make your hypothesis a theory. The explanatory power of a theory comes from the *many* lines of evidence that support it. And culturally controversial theories like evolution require even more support.

23. TW: April chose her language well: biological distinctiveness *may* intersect with humanity's theological uniqueness. That doesn't make them identical. Humanity's unique ability to blush isn't *necessarily* bound to our *imago dei*.

It may help to clarify how science works to appreciate the weightiness of a theory. In the sciences, what scientists accept as "true" is clearly defined by the practices of science, which can be simplified as predicting, making observations, conducting experiments, gathering facts and data, and forming explanations. Knowledge that emerges from these scientific practices is considered reliable because it has been subject to rigorous peer review, where experts scrutinize one another's scholarly work, research, or ideas *before* it can be published. Nobel Prizes in science go to researchers who make new discoveries or help to shift the direction of scientific thinking. If there were irrefutable evidence that could be replicated by numerous research teams that disproved biological evolution, this would be groundbreaking and Nobel Prize-worthy. But it's never been done. In fact, the past hundred years of research findings in anthropology, embryology, biogeography, genetics, paleontology, biochemistry, and comparative anatomy all support the claim that all life on Earth shares a common ancestor and has evolved over billions of years. This doesn't mean that the theory of evolution has not been modified as new evidence emerges. In fact, this is precisely how science is supposed to work; new evidence can lead to rejection of a theory, or adjustments to a theory. The model of an atom has shifted in such ways over the past century. Similarly, the biological theory of evolution has been modified, yet the foundations of the model still hold strong and are supported by numerous lines of evidence.

Oftentimes, the expression that evolution is "just a theory" tends to be motivated by a desire to question the reliability of claims about evolution. But evolution is not a siloed field of scientific discovery. The science behind the theory of biological evolution is not distinct or unique from all the other scientific discoveries our lives depend on every day. For example, the chemical and physical principles of chemotherapy are the same scientific principles that geologists use to date fossils and the age of Earth. To reject the findings of radiometric dating but accept the medical practice of chemotherapy is contradictory.[24] We trust in advancements from

24. Nichols and Penner, "Nuclear Chemistry and Medicine."

science every day, from the technology that put the words on this page or screen to the fabric of our clothes, the toothbrush and paste we use, and the medicines we take. Almost everything you interact with on a daily basis has its foundations in the claims and findings from scientific research. Nevertheless, science is not all-knowing; science has limitations.

The desire to question evolution can owe to a perceived threat: that affirming evolution will weaken confidence in God. But the tools of science cannot be used to answer questions about a creator or higher power; these questions require contributions from philosophy and theology. Another way to phrase this is that different types of questions demand different tool kits. The tool kit scientists use restricts the reality we can account for and establishes bounds on what we can know. The analogy of studying a flower can be helpful here. A flower grows from a plant and attracts pollinators; this is its "purpose." Using our eyes, a hand lens, or microscope, we can study the flower to learn why some insects are attracted to its shape (for instance, male bees can misinterpret part of an orchid flower for a female bee). Or we could use a refractometer to analyze the composition of the flower's nectar (e.g., the percent of fructose versus sucrose) because certain animals are attracted to particular compositions. Or we could capture the scent of the flower using a mass spectrometer to analyze those chemicals because certain animals are attracted to particular scents. Yet none of these methods fully explain why humans see flowers as appealing for aesthetic purposes, as expressing love, or as a token of sympathy. To understand all of these aspects of our attraction to flowers, we would have to use tools from the social sciences as well. We would need to understand cultural influences and historical traditions. So, to gain a full understanding of a flower and its appeal to animals (including us), we need to use different tools that provide different information from different angles. I like this simplistic way to explain it: you cannot use a microscope to understand poverty.

Science explains features of the world by describing the processes in nature that produced them, but science cannot make moral judgments based on those explanations, nor can it answer existential questions like the purpose of life.[25] While those existential questions might not be answerable by science, they are valid questions to which we desire answers. Sometimes those answers include supernatural explanations involving processes that occur outside of nature. Even though a few vocal scientists have chosen to criticize Christianity and claim that scientific data proves that there is no God, the majority of scientists are merely striving to make sense of the universe around them and propose coherent explanations for relevant evidence.[26]

Is evolution testable?

Absolutely! An essential part of how science is conducted involves developing a testable prediction. My earlier examples demonstrate how scientists test evolution predictions. In bacteria, researchers measured the rate mutations occur and predicted how many generations it would take for the population to evolve to resist a tenfold increase in antibiotics. As scientists searched for the origin of whales, they predicted anatomical traits that the intermediate species would have, and when intermediates were found, those predictions matched well with the fossil data. Similarly, when scientists hypothesized that early species moved from fresh water to salt water, they predicted that the oxygen molecules (isotopes) would reveal a matching shift, and the data support this claim. There is no end to the number of predictions scientists have made and tested over the past hundred years, but the bottom line is that the few examples I've provided in this chapter show that scientists can both observe evolution occurring on a small scale today, and

25. TW: This reminds me of Jonathan Sacks saying "science takes things apart to see how they work. Religion puts things together to see what they mean." See Sacks, "Rabbi Sacks on 'The Great Partnership.'"

26. TW: This is certainly what I see at my own Christian college and what our science faculty report about their colleagues practicing science in Christian as well as non-Christian settings. DE: This is my experience as well.

at the large scale, make predictions about intermediate species based on existing fossil data.

Human chromosome #2 offers a great example of the testability of evolution as well as providing robust genetic evidence that primates and humans share a common ancestor. Chromosomes are the strands of DNA that are encoded with genes. Scientists expected primates and humans to have the same number of chromosomes, but were surprised to find that great apes have forty-eight chromosomes (twenty-four pairs) while humans have only forty-six (twenty-three pairs). For years, scientists proposed two possible explanations: the ancestor for hominids had twenty-three pairs of chromosomes and primates carry a split chromosome, or the ancestor had twenty-four pairs and two chromosomes fused together in the human lineage. Near the end of the twentieth century, researchers discovered that at some point after the common ancestor for chimpanzees and humans, a mutation occurred such that two ancestral chromosomes fused together. The data show that human chromosome #2 has all of the genetic indicators of two separate chromosomes found in other primates: (1) the banding pattern of genes, (2) two centromeres, and (3) telomere sequences at both ends but also in the middle. I highly recommend a three-part YouTube video series on this topic by Christian geneticist Darrel Falk.[27]

While we may not be able to visually observe large-scale evolutionary changes occurring in our short lifetimes, evolution *is* testable. Researchers from various fields of science make testable predictions about evolution, perform experiments, examine DNA, or dig for fossil data, and if the data are verified and robust enough to pass peer review, their findings become part of the many independent lines of research confirming evolution.

27. Falk, "Chromosome 2 Part 1."

Why can't schools just avoid teaching about evolution?

I suspect that in the next decade, evolution will be in the news more than ever before because the science of evolution is becoming more and more relevant to our daily lives. For example, there is an emerging field in healthcare called *evolutionary medicine.* That means using the basic science of evolutionary biology to find ways to prevent and treat disease, and using studies of disease to advance basic evolutionary biology. Why do we have wisdom teeth? Why do our coronary arteries clog? Why do women have a narrow birth canal? Why do we have a food passage that crosses the windpipe? Questions like these can be explained if we understand how we evolved.[28] Evolutionary medicine also asks why ailments like back pain have been a problem for all hominin species since they first walked on two legs. Evolution explains the traits that leave us vulnerable to disease and other physiological challenges, as well as explaining why so many other aspects of the body work so well. Christians are already benefiting from this medical insight, and thanking God for the relief it brings. This is no more inconsistent than saying grace before consuming a meal of foods we evolved to crave. There does not need to be tension or inconsistency between evolution and faith.

Evolution also explains the increase in superbugs, or antibiotic-resistant bacteria. If we only kill off some of the harmful bacteria in our bodies, the resistant bacteria will be left behind to reproduce and pass on that resistance. To combat these ever increasing superbugs, we as consumers of medications need to understand the importance of taking antibiotics only when absolutely necessary and following our doctor's orders when we do take them. And as I write this chapter, we are nearing the end of year two of the COVID-19 pandemic. Just this past month, scientists

28. TW: There are therapeutic applications too, in developmental and abnormal psychology, with approaches that have already proven fruitful and are already widely used in Christian circles.

discovered a new, less virulent strain of the virus. Understanding the evolution of bacteria and viruses is critical to our livelihood.[29]

A third example of evolution's relevance to our lives focuses on the damaging effects humans are having on the environment. Researchers find that ecological changes to an environment, often due to human actions, can have major consequences for organisms and the functioning of ecosystems, and can often lead indirectly to harmful effects for humans. New findings are revealing that disturbed environments (e.g., deforestation and land development) can sometimes lead to rapid evolution among some species in the affected community, and we don't know yet what the long-term effects will be. What we do know is that evolution can be a game changer even in the short term, which has important implications in areas such as disease and pest control, conservation and fisheries management, and decreasing harvest yields.

My final example is a more personal one. About fifteen years ago, I became interested in the ways clergy manage the topic of evolution after an experience my son had at his junior high church camp. At the end of a long weekend at camp, he came home excited to tell me all that he had learned about the evils of evolution from the youth pastor. For half an hour he shared with me what this guest pastor preached about during the three-day camp. My son fully bought into the idea that as Christians, we have to reject evolution. After listening to him intently while quietly agonizing over each sentence he spoke, I very gently began telling him that this was not what I thought about evolution. I proceeded to share a bit about how I reconcile evolution with my faith, but he began to cry. In retrospect, I can see that he was extremely distraught that two authorities in his life, his pastor and his mother, held opposing positions. It was then that I realized the challenge our youth face when clergy say one thing and other authoritative figures (like

29. TW: The discourse around COVID-19 also shows how widespread and assumed that knowledge of evolution is in our culture. Christians who remain ignorant of it or rely on caricatures turn themselves into little fragile metaphorical island populations that neither represent the kingdom's mission to reach all nations, nor thrive in the long run. Douglas's chapter addresses this aspect of our calling.

teachers, parents, or the news media) say something else. Youth can get very confused and feel trapped or overly conflicted, and then transfer this frustration to their view about religion or worse yet, their belief in God. My son actually wrestled with the evolution issue for a couple of years before he came to accept and reconcile the science with his faith. His distress could have been avoided if the Christian community had a more open and conversational approach toward evolution.[30]

The bottom line is that the science of evolution is not going away. It's the governing paradigm for the whole modern field of biology, a theory that has much explanatory power for both human health and the health of all life on Earth. Instead of avoiding evolution because it is perceived as too contentious, we instead must do the difficult work of thinking about our scientific, philosophical, and theological assumptions about evolution and turn negative emotional responses into opportunities for dialogue and learning.

Does evolution show me things that make a difference for how we should understand God or God's creation?

Definitely. My view of who God is expands with each new scientific finding I learn about. So much of what science has discovered, from the intricate details of human development to the enormity of the universe, confirms for me the magnitude of God's creativity and power. Just thinking about an expanding universe boggles my mind. Expanding into what? What was there before it expanded? How could God possibly create such complexity from the molecular world to the vastness of our universe? Any narrow view of God and how God created that I once held has been blown apart by

30. TW: Christians told April's son the same thing she had heard from her science teachers: that affirming evolution meant abandoning biblical faith. He was blessed to have support to help him work at his own pace through the tension between two traditions and relationships he respected. This book's authors hope you can be part of such a support network. How many believers turn their backs on science, and how many fans of science turn their backs on Jesus Christ, because they regard the two as incompatible, and lack that kind of support to help them work through the issue well?

what I've learned from scientific discoveries of the past century. I value the findings from science because they cultivate my humility; we really cannot grasp the ingenuity, brilliance, and grandeur of our Creator.

I enjoy helping my students cultivate a sense of awe and wonder at God's creative processes, and helping them see that science is not a threat to Christianity, but rather an opportunity to expand our view of God. Midsemester in my biology course I start one class session with the lights dimmed, a song playing over the speakers, and beautiful images of plants, animals, and the universe scrolling on the screen. After fostering a sense of awe visually, I take a few minutes to pause with the class and verbalize that science helps us to better understand the glory and majesty of God through God's creation. This is just one of the many times we talk about the majesty of God. Instead of threatening their faith, many of my students begin to view science as a lens into the magnificence of God's creation. Science allows them to understand *how* God created. Susan, a sophomore biology major, said "science is so intricate with genetics and cellular functions, for me it just glorifies what God can do. [My friends and I] would start talking about science and we would say God is so cool." When Caden, a fourth-year chemistry major, spoke about reconciling science and faith, he told me, "I think, 'Isn't that amazing?' Just acknowledging for a moment that, 'Wow, this is our creation! This is what our God has done and created everything. He's allowed us to explore it.'"

Most of the science majors at my university marvel at God as creator and author of the universe. This contrasts with the misconception that learning about evolution and other scientific discoveries diminishes or threatens one's faith. In fact, I've conducted a few studies to show that this claim is simply not true. When students have a path to reconcile consensus science with their faith, they will boldly pursue scientific truth wherever it may lead while maintaining their faith. If everything is part of God's creation, then there is no need to fear science because all truth, ultimately, will lead us to marvel at the God who created.

For further reading:

BioLogos.org.
Darrel Falk, *Coming to Peace with Science.*
Gregg Davidson, *Friend of Science, Friend of Faith.*

2

A Pastoral-Biblical Perspective

Douglas Estes

Welcome to this part of the conversation! Before we begin, three starting points for our discussion.

First, the purpose of this chapter is neither to convince you of nor convert you to an evolutionary perspective. Although I accept the observational data and research that evolutionary biology and adjacent fields have brought to light in the last two hundred years, this data and research does not—and for the most part cannot—give insight or answers into the most important questions that people face: Is there a God? Is it possible to know God? How can I know that I know God? Therefore, my personal interests are primarily to help people understand who God is, how they can live a blessed life here on earth, and how they can be as sure as anything that there is an eternal home for them once they pass from this world. Part of living a blessed life on earth, though, requires that we learn to navigate the world in which we live. I believe it is essential that Christians understand evolution—the science of evolution, not caricatures—in order to navigate their world well.

Second, throughout this chapter I use the word *evolution* to mean the sum and total of the observational data and research that provides a scientific basis to explain how life unfolded (and continues to unfold) on our planet. I do not use the word *evolution* as people most often use it in our culture: the naturalistic and nihilistic view that everything came from nothing and therefore there is no meaning to life other than what we make it (sometimes referred to as *evolutionism, Darwinism, naturalistic evolution,* or a type of *scientism*).[1] In fact, as we will learn throughout this book, there are many reasons to reject the latter even if we accept the former. When we talk about evolution as people often mean it in our culture today, I will write it as "evolution" (using scare quotes) so that we will know I mean it in a popular sense and not in its scientific sense. This is part of the reason that I do not self-identify as an evolutionist. Although I accept the data and appreciate the explanatory power that evolutionary biology has attained, it is still a very young field of study (hardly 150 years old), and I believe there is much, much more to be studied and learned before we can fully appreciate how the world that God created works through observational data.

Third, although my background is in the Bible, you won't read me citing much Scripture throughout this chapter. The reason is to avoid any pretense of prooftexting.[2] Not only is prooftexting bad, but it is a common trap Christians have fallen into when discussing science. The primary purpose of God's word is to reveal who he is and who Jesus is and who we are—and how we can all get (back) on the same page. Of secondary issue within Scripture are all the variables that life throws at us on a regular basis—for example, what to do when anger strikes (James 1:19–20). These are variables that Scripture cannot cover in every detail *ad infinitum*. Instead, it offers some parameters and examples and reminds us this is why we have prayer, the Holy Spirit, and the church. In comparison, I

1. Haarsma and Haarsma, *Origins,* 180.

2. To *prooftext* is to take a verse from the Bible and apply it to a life scenario that is out of context from its original use. This is a very common approach to biblical interpretation—for examples, just look at most social media posts that cite a Bible verse, or listen to the speech of any politician—but it is one that ranges from weak to deceptive interpretation. We use this term today to describe a misuse of the Bible. For further discussion see Kaiser and Silva, *Introduction,* 31–32.

consider evolution a tertiary issue, albeit an important one. Although the Bible talks about creation and the world in which we live, it does not talk about our world in a scientific way—science is a modern way of understanding the world that, while immensely valuable, did not exist when the Bible was written. Personally, I believe the Bible is the inspired word of God and is infallible in its revelation of God to people.[3] I also believe God has done something unfathomable in his creative work in our world, something that humans cannot (and probably will not) fully understand, but we can still observe the world from our perspective and create meaningful theories that are explanatory and effective.

With these ideas in mind, let's turn to how we address evolution and faith:

Can a person believe in evolution and believe in Christ (be saved)?

Yes. One can believe in evolution and be a fully committed believer in Jesus.[4] Let's consider this from both sides of the question.

If evolution is true, then we should acknowledge it because all truth is God's truth. What this means is that anything that we learn in our world that is true ultimately originates with God. As John Calvin famously wrote, "All truth is from God; and consequently, if wicked men have said anything that is true and just, we ought

---

3. *Infallible* means "incapable of failure" or "unable to err." Many Christians use this concept to describe the Bible. Although the idea is not explicitly stated in the Bible, it is implied, and it is the logical conclusion of the Bible's claim to inspiration. Since the Bible is inspired by God, and God does not lie or mislead, the implication is that God's word would not lie or mislead; thus, the Bible is unable to err in what it reveals to people. Infallibility does not preclude erroneous interpretation, however, as we will discuss below.

4. AC: Scientists don't like the phrase "believe" in evolution. There are various reasons why, and some are explained in Branch, "What's Wrong with 'Belief in Evolution'?" For me, the primary reason is that if I say I "believe" in evolution, then this could lead others to think that I cannot cite evidence to establish my position. Since I can provide an abundance of evidence, "belief" feels like the wrong term. I "accept" evolution.

not to reject it; for it has come from God."[5] Therefore, if evolution is true—regardless of how it is promoted, packaged, or sold—then we as Christians should embrace it because it is true. (We'll discuss the packaging part below.)

I also do not believe there are any inconsequential truths. If it is true that you know how to read, then there is a reason for your knowledge of reading (not to read this book, but to read God's word!—and to learn and to communicate well with others). God has a *plan* that includes your knowledge of reading. God also has a plan that includes all truth, from 2 + 2 = 4 to understanding black holes and top quarks to discovering the observational data that demonstrate evolutionary theory. God even has a plan for dinosaurs and viruses. We may not understand what God's plan is for evolution, dinosaurs, or black holes (or perhaps even why you are here reading this book right now), but if we trust in God then we trust that he has a plan for what he does.[6] As Christians, we are not required to understand much of God's plan—God's ways are higher than our ways (Isaiah 55:8–9)—but we are expected to live what we do understand as best as we can. Thus, if we know something is true, or think it is likely true, then we should recognize that God had a reason for this truth even if we don't understand it.

However, *believe* is a funny word here, and it illustrates one of the major challenges in talking about evolution. If we read the Bible, we find that Jesus asks us to believe in him (John 6:29). In this sense, the word *believe* means something closer to "commit your life" than "understand as true." For example, to explain Jesus to others, John the Baptist argued "whoever believes in the Son has eternal life, but whoever rejects the Son will not see life" (John 3:36 NIV). Although we can reject knowledge, it is not simply the knowledge about Jesus that John means when he warns of rejecting Jesus—it is the failure to fully and completely commit that leads to a rejection of the Son. To put it in slightly different terms,

5. Calvin, *Commentaries*, 300–301. Likely Calvin was echoing Augustine, *Teaching Christianity* 2.18.28.

6. My biblical reference for this is Genesis to Revelation; all of the Bible reveals the momentum of God's action in our world to draw people to his Son.

a person can understand that evolution is true and not reject Jesus. Likewise, a person can accept that evolution is true and believe (commit to *and* understand as true) the Bible—especially if they are willing to admit that they don't have all the answers about how the two fit together in every situation and scenario. This is because in any Christian view of creation, there is both natural activity of God and special activity of God; it's getting the right balance between when God did what and how that is the hard part.

Evolution does not ask anyone to believe (it can't, it's just a theory that explains the data!). But Jesus does more than just ask us to acknowledge him; he asks us to commit all of who we are to him. This is why Paul writes "if you declare with your mouth, 'Jesus is Lord,' and believe in your heart that God raised him from the dead, you will be saved" (Romans 10:9 NIV). What Paul tries to say is that both externally and internally we need to fully embrace that Jesus is the Messiah. If we can do this, we are now children of God—it has no bearing on whether we understand or acknowledge scientific research. Whether or not the knowledge we have about the world is accurate or inaccurate does not limit God's ability and willingness to save us.

I argue for this because I am always concerned that we Christians neither add to nor take away from how a person can have relationship with God. I find that Christians often implicitly add to salvation by making statements that imply that if some theological position is not held, or worse some information not understood, then this person may not really be a Christian. This is problematic because the Bible seems pretty clear on how we are to judge the faith of people: we must consider the fruit they bear (Matthew 7:15–23). In the New Testament, "fruit" seems to refer to two things: being on message with God and acting in ways that support that message. It doesn't seem to speak to a knowledge base beyond the basics. This is why Jesus suggests that those who come into the kingdom are like children: Children can be fully on message with God and be obedient to his requests (bear fruit), even if they don't understand everything (Matthew 18:3).

If I accept evolution, and in the end it isn't true, will I be judged unfaithful? Or seen as one who has fallen away?

No. A person's salvation is tied to whether or not they have put their faith in Jesus. It is not tied to a specific base of knowledge other than knowing who Jesus is and what he does (cf. John 10:38). Furthermore, lacking in knowledge is not itself a sin.

I believe this to be true because I know people who for physiological reasons cannot comprehend either creationism or evolutionary theory yet believe that Jesus loves them. (Probably you know someone like this as well.) This is actually true of all people, because even the smartest people still have limited reasoning abilities and knowledge retention. If we look back through history, we will see many well-known Christians who had erroneous views on a whole host of issues including the origins of life, and beyond. This is to say nothing of the millions of Christians who lived who were not privy to the latest developments or who lived in less educated societies. I know that I have inaccurate and erroneous views on science (and beyond). If someone reads this book in the twenty-third or twenty-fourth century, I am sure that there are some cultural, scientific, or even biblical interpretation views that these future readers would consider flawed (just as we find some ideas from centuries ago flawed). Fortunately, God will not prevent you or me a place in his house simply because we have "human ways" that lead to faulty conclusions about our world.

From a historical perspective, the early church wrestled with what was needed to best explain faith to people. One of the great achievements of this discussion was the Apostles' Creed (as well as other creeds such as the Nicene Creed). If we read the creed, it tells us what we need to believe, but it also tells us what we need to know to have faith. We don't need to have a perfect understanding of creation, the diversity of life, Adam and Eve, or heaven and hell. In fact, most of these concepts are absent; the few that are present (e.g., God is the maker of the world) are meant to teach us about God, not help us to understand these concepts in abstract.[7] Grant-

---

7. I find that one struggle Christians always face is the temptation to add to

ed, if we have a skewed perspective on some of these issues—such as if we believe that Adam and Eve were extraterrestrials—it could lead us to a crisis in our faith, but not necessarily a loss of true faith. The difference, of course, is that while there is no evidence that Adam and Eve were extraterrestrials, there is significant and substantial evidence for evolution. If we can observe physical evidence that reveals the way the world works, whether the cycle of tides or the evolution of coronaviruses, this evidence on its own cannot cause a loss of true faith—even if we may have much work to do to understand what we see. It is our assumptions and misunderstandings *about the world* that lead to loss of faith, not the evidences we find *in our world.*[8]

Although this may seem a pragmatic argument, Jesus was at times pragmatic also. One of my favorite statements in the Bible comes when Jesus resorts to what we would today call a fully pragmatic argument: "Do not believe me unless I do the works of my Father. But if I do them, even though you do not believe me, believe the works, that you may know and understand that the Father is in me, and I in the Father" (John 10:37–39 NIV). Here Jesus lowers the threshold of faith so that even if people don't believe he is telling the truth, they can just look at his actions, and still believe. Jesus invites these people to come to faith even with a

---

and subtract from the parameters of salvation. Often, well-meaning Christians will claim that to be a Christian one has to agree with their interpretation on topics such as creation, heaven and hell, the role of the Holy Spirit, or the nature of the church. While these are important issues, they fall under the sphere of discipleship and sanctification (as God's children we learn about his works and his plans), not salvation.

8. History is full of examples of faithful people who had poor or limited understanding of the physical world. For example, all ancient people had an erroneous view of how the human body works—notably how we see and how we feel emotions. This is true and will remain true as human knowledge progresses. For example, I cringe whenever I watch a movie that depicts medical standards from a century ago; they were incredibly crude by today's standards. Each generation believes it knows everything that can be known; but as time progresses, humans will continue to sharpen their knowledge of the world. I am positive that many of the "scientific" ideas I have today will prove laughable to my great-great-grandchildren. Fortunately, the many erroneous ideas I have now are unrelated to the value of my faith.

terribly erroneous belief—that Jesus wasn't telling the truth! Conversely, even if we believe the consensus view of scientists in our age (on topics such as evolution), and that consensus view later turns out to be wrong, that is hardly as scandalous as being wrong about Jesus. We can be wrong about earthly things and still believe.

What about those who come from a strict creationist background but see some value in evolution also? What do we do if we just can't decide?

Keep learning, and keep asking thoughtful questions about both science and interpreting the Bible and theology. I don't say that as a dodge: I am in no way implying that if you learn more (hear: get smarter) that you will suddenly have a conversion experience and see the Darwinian light. This is often what you hear from proponents of scientism.[9] I don't think you have to decide, certainly not in any rigid way. It is fine to have more questions than answers. I do suggest that you keep learning because the more you understand what science says (and doesn't say), the more it will give insight into how you read the Bible (specifically the parts that allude to our natural world), and the more you study the Bible, the greater wisdom it will give you in discerning where evolution stops and "evolution" begins.

Let's consider the learning aspect for a moment as a key value in learning about evolution. The sciences are every bit of a calling as other fields that have received that label in the past—public service, law, or medicine, for example. Not all of us are called to be biologists (like April). Among those who are, nearly all accept evolution as the best explanation for the evidence they see when

9. Scientism is "the belief that science and its methods provide the only fully valid route to gaining knowledge and for answering questions, to the exclusion of other methods and disciplines," as defined in Principe, "Scientism and the Religion of Science," 42. In the sense that people use the word culturally, it suggests that science has (or will have) all of the answers to life's questions. Some philosophers and scientists have attempted to rehabilitate the idea, making it more neutral—science can provide some answers to all of life's questions; for an example see Peels, "Conceptual Map of Scientism," 30.

they study the world around them. Among those there are many (like April) who have expertise in the natural world and who are believers. Their testimony of how God works in evolutionary ways in our natural world is worth thoughtful consideration. If we have a chance to listen to scientists like April, it will only naturally make our understanding of science richer, and by extension, our understanding of the Bible more accurate.

An observation that is sometimes missing from these discussions: Almost all scientists are regular people who merely investigate the data in their narrow fields and write up their conclusions in very, very narrow areas of research (or they implement the results of this research). They do not go around making grand pronouncements about science, much less the implications for science as it relates to God. This is lost because our media often hype the few yet vocal scientists who are critics of the Christian faith under the guise of science because it creates controversy and therefore generates revenue. But this hype is not science.

I was taught that evolution is false.
Why should I learn about evolution?

There are several major reasons why we should learn about evolution even if we are not fully convinced. I will mention three here: greater understanding, shrewdness, and cultural engagement.

One important reason we should learn from science about evolution is to possess an overall greater understanding about the world in which we live. As Paul observes, God reveals himself both in direct communication with humans (mostly in very special situations) and through the world in which we live (Romans 1:20). In fact, Paul explains that we can clearly see features of God in the world around us. Therefore, if there are a group of people whose calling is to observe the world around us in great detail (let's call these people scientists), then we can learn a great deal from those people (even if they don't have all the answers and can't always reconcile every theory, as in physics).

From a practical perspective, we should learn about evolution to avoid having our understanding of evolution be a caricature of the actual theory and data. In college, I majored in chemistry. Since chemistry was a very small department, I am almost certain all of my professors knew I was a Christian. I wrote a Christian column in our secular college newspaper; and some of the other chem majors would make harmless "God squad" jokes about me and my faith before or during class if there was a religion angle. None of the faculty ever made an issue of it. A few of the chemistry faculty attended church regularly. One faculty member in an upper-level lab confided to me that because of life experiences he was wrestling with the whole God thing and was thinking of taking the next step and going back to church. Maybe some of the faculty didn't think much of my Christian faith, but if so, I never detected it even in working closely with almost all of them in one class or another. However, when I went to seminary, I encountered a number of caricatures about both science and scientists among not just students but faculty that crept into class discussions.[10] Since all truth is God's truth, if we discount, or worse, mock, something that has truth in it, then we dishonor ourselves before others and run the risk of dishonoring how God creates in our world. Again, I don't believe God calls most people to know for certain whether evolution is the most accurate theory of life origins, but he does call us to be accurate in our dealings with other people, including people who do not agree with us.

The second reason we should learn about evolution is so that we can exhibit a shrewdness about how we navigate the world that we live. Using a parable, Jesus questions why worldly people are more shrewd than godly people (Luke 16:8).[11] Make no mistake,

10. AC: I can speak from personal experience that in a church environment, when a scientist hears these kinds of statements from pastors or congregants, the scientist is reticent to return to church, or even acknowledge belief in God if it means leaving our science—our calling—at the door.

11. Readers consider this parable to be one of the most difficult in the Synoptic Gospels. Because ancient versions of the Bible did not have punctuation, modern versions sometimes must guess whether a sentence is a statement, question, or exclamation. Here I find it likely that Jesus uses a question to push

the majority of the people that we speak with do not fully understand evolution; most, at best, know more about "evolution" (the
cultural myth) and have only a rudimentary understanding of the
theory that they fit into their assumptions about life (often their
view on whether there is a God or not). Because our culture accepts "evolution," it does little good to dispute a topic that is likely
not well understood (or misunderstood) by most people and of
tertiary importance to what the Bible teaches about God (fascinating, important, but still tertiary). All truths are God's truth, but not
all debates are meaningful, and not all hills are ones to die on. This
is why I believe the best thing you can do is keep learning about
evolution (to whatever degree you are led) so as to be effective in
navigating conversations with others.

During our collaboration, Telford reminded me of a similar
argument from Augustine, which we felt would be worth quoting
at length:

> Usually, even a non-Christian knows something about
> the earth, the heavens, and the other elements of this
> world, about the motion and orbit of the stars and even
> their size and relative positions, about the predictable
> eclipses of the sun and moon, the cycles of the years and
> the seasons, about the kinds of animals, shrubs, stones,
> and so forth, and this knowledge he holds to as being
> certain from reason and experience. Now, it is a disgrace-
> ful and dangerous thing for an infidel to hear a Chris-
> tian, presumably giving the meaning of Holy Scripture,
> talking nonsense on these topics; and we should take
> all means to prevent such an embarrassing situation, in
> which people show up vast ignorance in a Christian and
> laugh it to scorn. The shame is not so much that an ig-
> norant individual is derided, but that people outside the
> household of the faith think our sacred writers held such
> opinions, and, to the great loss of those for whose sal-
> vation we toil, the writers of our Scripture are criticized
> and rejected as unlearned men. If they find a Christian
> mistaken in a field which they themselves know well and

the disciples to admit the problem of a lack of shrewdness on the part of believers (cf. Psalm 18:26).

hear him maintaining his foolish opinions about our books, how are they going to believe those books in matters concerning the resurrection of the dead, the hope of eternal life, and the kingdom of heaven, when they think their pages are full of falsehoods on facts which they themselves have learnt from experience and the light of reason? Reckless and incompetent expounders of Holy Scripture bring untold trouble and sorrow on their wiser brethren when they are caught in one of their mischievous false opinions and are taken to task by those who are not bound by the authority of our sacred books. For then, to defend their utterly foolish and obviously untrue statements, they will try to call upon Holy Scripture for proof and even recite from memory many passages which they think support their position, although *they understand neither what they say nor the things about which they make assertion.*[12]

The final reason to learn about evolution is to be better equipped at cultural engagement. Even if we are not fully convinced of evolution, learning about it from scientists like April will equip us to navigate and minister in our culture. It is a fact that our *culture* accepts that evolution is true. Therefore, if we don't understand evolution well, and reject it, it will limit our ability to navigate our culture. For example, our *culture* accepts that US astronauts landed on the moon in 1969. Our *culture* accepts that the speed of light is constant. Our *culture* accepts that there are three phases of matter. Our *culture* accepts that the sum of the angles of a triangle is 180 degrees. Now, the first is accurate, the second is accurate as far as all evidence suggests, the third is an oversimplification (there are anywhere between four and twenty-six phases), and the fourth is inaccurate as it is based on an assumption (180 degrees is the sum of all angles of a triangle in two-dimensional space, but not necessarily a triangle in three-dimensional or other space).[13] However, it is hard to navigate culture if we do not accept—or at minimum are unaware of—what culture accepts.

12. Augustine, *Literal Meaning of Genesis*, 42–43.
13. The last is a well-known thought experiment.

If the average person you meet outside of your Christian sphere believes in "evolution," then you can't engage well with those people without having some knowledge of the topic. Now, some will think of the old saw about the youth pastor who, to be able to discuss drunkenness with his students, got drunk frequently to gain the experience to talk to his students. But that is not this, because there is no sinfulness in gaining knowledge of human theories. There are many examples of this in the New Testament though they are not always obvious to modern readers. For example, we know that Jesus understood human medical theories even though he was aware they were not effective and he did not need them himself to heal. At the same time, Jesus chose to use ineffective healing practices to engage the people around him (John 9:6–7). If Jesus can understand, and make use of, erroneous ideas about medicine, we can certainly understand and make use of far more accurate ideas about evolution. Even more so, awareness of the beliefs of people around them is exactly what Paul does—Paul often is aware of his day's predominantly Greek philosophy and uses it to engage the culture around him (Acts 17:16–34). Thus, if Paul could learn—and sometimes accept—Greek philosophy to engage people, we too can learn—and often accept—scientific theories. What is interesting here is that these Athenians had an inaccurate view of God, even though a more accurate view of God was available; in the same way, there are many in our world who view the world inaccurately through "evolution" even as a more accurate view of our world (both scientifically and theologically!) is available to them.

I do not want this argument to be overly cynical, and to make learning merely a means to an end. Instead, a by-product of greater knowledge about the way God reveals himself in our world is a greater ability to minister to others—especially the ability to share the gospel with others. A great example of this in my own life occurred when I was younger and met people working in computer science and information technology. Although not advanced, my knowledge in these areas is greater than the average layperson's. My ability to "speak their language" just as Paul spoke the language of the philosophers in the Areopagus created a great number of opportunities to share the gospel in a way that helped it to be received by those listening.

This is the other part of the reason why, as I noted earlier, I do not self-identify as an evolutionist (the first part being the cultural misunderstanding of what evolution is). When people ask me what I believe (*believe* being the pesky word again), I tell them "it's complicated," and if there's time I acknowledge what we can know about God, what I perceive our limitations are in biblical interpretation, and then what we can know about the natural world, and what our limitations are there. I find that if we are immediately against what culture accepts to be true, it makes it harder for us to navigate (or it is a hill we should be prepared to die on); but if we are still thinking through it, then we learn much more and are able to have spiritual conversations with those who don't agree with us much more easily. I find that most people are reasonable about this, including most people who work in the sciences. This is because most people who work in the sciences are not on the forefront of evolutionary biology; they accept the theory, but even they are not always aware of all the details and how it all fits together.

Where I see this play out in my life recently is in conversations with scientists about astrobiology. As of writing this, there is no concrete evidence that there is life on other planets. However, there is circumstantial evidence that life may be on other planets. When and if we discover that life, it is likely that it will not be the walking-and-talking kind. Instead, it will most likely be the microscopic kind. And the "is this really life?" kind. If this happens, it will trigger a massive debate in our world about the origin and meaning of life. Although I tend to accept there is life off of Earth (as defined by science, not necessarily recognizable as such to regular people), in conversations with astrobiologists I still lead with questions, not my opinions. It is always well received.

Why does "evolution" mean different things to different people?

"What are you reading?" the flight attendant asked me. I was stuck in the dreaded middle seat, and, trying to make do with an uncomfortable situation, held the book I was reading up a little higher than normal.

I turned the book so she could read the cover.

"I thought so! Well, good for you! I wish more people were more enlightened."

The book I was reading was a book on evolution by a well-known anti-God evolutionary biologist and pop philosopher.

After some brief chitchat, I admitted that while I was interested in the science of the book, the book itself was of limited value because the goal of the author was less educating the reader and more scoring (cheap) points against people who believe in God. When the flight attendant realized that I was a Christian, it signaled the end of her assumption of my enlightenment—and the conversation.

I believe this story is instructive about the way "evolution" functions in our society. I was reading the book because I wanted to learn more about evolution. Yet it is obvious that the author uses his knowledge of evolution as a vehicle to introduce his personal beliefs about the universe. That's fair; there are Christians who do the same. However, it is unfortunate because this activity also discourages some Christians from listening to and thinking about evolution. It also creates a tribal flag for other people. This is why the flight attendant made several assumptions about me merely from my reading the book. While one can make assumptions about me no matter what I read, the debate has entered culture to such a degree that the flight attendant naturally believed that I believed in "evolution"—and therefore assumed that I did not believe in God. When I say "debate," I don't really mean the debate between evolution and Christianity, I mean the debate between "evolution" and theism, with all the tribalism that comes with it.[14]

14. TW: This kind of tribalism makes mission a cross-cultural enterprise.

As I explained earlier, the problem is that there is evolution and "evolution." Even among evolutionists, there are multiple proposals for how God works (or does not work) through evolution.[15] As Christians navigating the twenty-first century, I believe we need to understand and be prepared to explain our faith to both—people who accept evolution as well as people who believe in "evolution." Sometimes those groups overlap, though I find that is true less than people expect. Only a little more than a third of people in the US graduate from college,[16] and of those college graduates, only about 6 percent major in biology or related area.[17] Not discounting self-learners, the percentage of people walking around with a reasonably sophisticated understanding of evolution is very limited. At the same time, most Westerners have had a dose of evolution (and often "evolution") in some "Biology 101" type of class either in high school or college. All this to say that these ideas—evolutionary theory and naturalism, probably deism, materialism, and a pinch of nihilism, too—are mixed together for most people in Western countries in a big philosophical stew that encompasses a little bit of evolution and whole lot of "evolution."

If we read about the history of the scientific study of evolution, we find very quickly that adjacent ideologies quickly began to attach themselves to the theory. When people began to understand the great explanatory power of evolution, proponents of naturalism (the world is all there is) and materialism (everything is just matter) were quick to read their ideologies into evolutionary

---

Missionaries to other tribes, even secular American ones, need to learn to transcend barriers and "translate" the gospel into terms their audiences can understand. Here Douglas is demonstrating the kind of investment and risky culture crossing that helps mission succeed.

15. Rau, *Mapping the Origins Debate*, 31–56. AC: Douglas initially used the term *theory* here in an early draft of this chapter. I commented that this is not an accurate use of the term, at least for a book that discusses science. I discuss the meaning of theory in my chapter, but this just reminds us that different groups of people use words differently, and it really matters when we try to fully understand something.

16. US Census Bureau, "Educational Attainment."

17. National Center for Education Statistics, *Digest*, column 19.

discussions. Some people also read deism (there is a God who created but is now absent) into the mix. To be fair, theists (people who believe in God) did much the same thing with the discovery of the big bang, the idea that the universe began as a tiny point and exploded outward continuing through today.[18] However, theism is vastly superior to naturalism and materialism if for no other reason than it provides hope and purpose to people that the other two ideologies do not. That's the weaker, pragmatic view that I find works best when building bridges with people who have little or no faith. Theologians like Telford make the stronger argument that the reason why theism is vastly superior to naturalism and materialism is because naturalism and materialism cannot explain things that don't fit the naturalist or materialist view of the world—most notably the resurrection of Jesus the Messiah—that demonstrate that God is real.

While evolution does mean different things to different people, I believe a shrewd Christian who is called to interact with adults in the Western world will need to be aware of the various views on evolution and "evolution" and how those views shape the people around them. Or to put it another way, Christians should learn about evolution not to debate evolution with other people, but to understand the various ideas about "evolution" that shape the modern individual. This will mean reading proponents of both evolution and "evolution." In doing so, it will prove to make us a far more effective witness to others as well as put us in a place where we are always growing through our interactions with others. It will also help us to show God's love working in us for others to see (Proverbs 12:26; Hebrews 12:14).

18. TW: Even if Christians win some of these battles, the battles themselves train us to see Christianity's credibility standing or falling on grounds that are foreign to its roots, and that turns Christianity into something it originally wasn't. Similar distortions happened in the long Christian engagement with classical philosophy. The claims of the apostolic faith are fundamentally historical in character, but many now treat it as basically a philosophy or a "religion," and thus an (often inferior) alternative to scientific knowledge.

Is it possible to accept some claims of evolution without accepting all claims? For example, if evolution is accurate, what do we do about Adam and Eve?

Yes. Although evolutionary biology is a young field, it is also a large field built upon many, many observations and studies. When we consider the theory itself, it is very possible to agree with many studies but not agree with all studies. This is in fact what scientists do themselves—they evaluate studies and those they judge to go against their own understanding or research they critique or challenge. If a scientist publishes a new study in a peer-reviewed journal, it does not automatically mean the study is correct or that the study will stand up to further scrutiny beyond the first review. As a result, scientists have not reached a final conclusion in all areas or that answer all questions within evolutionary biology. One example—from one end of the spectrum—where there is universal consensus among both scientists and nonscientists is microevolution. Virtually everyone agrees with this, in part because it is one of the earliest and most demonstrable arguments within evolution. Its arguments and research even predate Charles Darwin. One example—from another end of the spectrum—where it seems to me to be reasonable to be uncertain is in the ongoing discussion of human origins. Much of the research into how humans arose is still rather young, and there have been some hiccups and surprises along the way (such as Denisovans, one species of early hominid).[19]

Of course, this is not what most people mean when this question arises; most people want to know if they can cut out parts of "evolution," eliminating the parts that seem to go against God or Scripture. From that perspective, absolutely; naturalistic, materialistic, or deistic arguments that some scientists and much of pop

19. The Denisovans are surprising because they "point to a stunning diversity of hominins in ancient Asia," and are causing scientists to reevaluate their ideas on the rise of ancient humans; see Wei-Haas, "Multiple Lines of Mysterious Ancient Humans Interbred with Us." AC: To clarify, scientists are *not* debating whether humans evolved. My chapter provides survey results that show some 97 percent of scientists accept human evolution. Rather, it's the exact lineage's details, the when and what, that are still being uncovered.

culture embed within presentations of "evolution" can and should be rejected.

I believe this is where humility and honesty should come to the fore. There is much that science cannot yet explain about the world in which we live. That does not mean science will never be able to explain it, but it does mean that there are a great deal of unknowns. The same is true of the Bible—even though it is God's inspired word, there are a lot of questions we ask that it doesn't try to answer. There are also a lot of statements in the Bible that we don't fully understand (for example, every statement about God). When we look at our world through science or the Bible, we see more questions than answers. I believe that God created such complexity in our world both to astound us and amaze us. Why would an awesome God design a simplistic world that even his creatures could fully comprehend in a short period of time? Similarly, why would an awesome God design a simplistic Bible that even his creatures could fully comprehend in a short period of time?

The issue of Adam and Eve is instructive. I believe there was an Adam and Eve. I believe this because, among other things, this seems the best reading of the biblical passages that discuss Adam and Eve (even if we set aside many of the interpretive difficulties of who they really were) *and* because this seems compatible with observational data about human origins. That doesn't mean, though, that I accept all of the romanced views of Adam and Eve that have percolated through culture over the last few thousand years. I'm sure if we had video footage of Adam and Eve it would shock our modern senses on many levels. Recent pronouncements from genetics and populations studies (trumpeted by the media) have cast doubt on a historical Adam and Eve.[20] Yet there are theories that show that Adam and Eve could exist within the confines of evolution.[21] This issue raises more questions than we can answer

20. AC: A nice brief primer on a variety of views Christians hold about Adam and Eve is BioLogos.org, "Were Adam and Eve Historical Figures?"

21. To mention but a few examples: Swamidass, *Genealogical Adam and Eve*; Bonnette, "Rational Credibility of a Literal Adam and Eve"; and Applegate, "Why I Think Adam was a Real Person in History."

from *two* fields of study: biblical interpretation and evolution.[22] What does the Bible actually say (and not say) about Adam and Eve? And what can evolutionary biologists observe and not observe about a first (or special) pair of humans? My point is simply that there is a great deal of work ahead of us, and in the meantime, it is reasonable to be open to new data and new possibilities even as we take seriously the groundwork that scientists (and biblical scholars) have already established.

Every few years it seems there is a big scientific announcement that casts doubt on God. Why should I be positive about science?

Just as there is evolution and "evolution," there is also science and "science" (essentially scientism). What we hear trumpeted in the media is often "science," from people who use scientific research to draw philosophical and theological conclusions about the world— conclusions that are generally not entailed by the research itself. They are opinions masquerading as facts. In this, there is nothing new under the sun.

As Christians, we should be positive about science, even as we reject "science." When I was a pastor, I once met a man who worked in the sciences who was married to a woman who attended our church. He was a self-declared "evolutionary" atheist who peacocked around, feeling it part of his purpose in life to use his science background to discourage people from believing in God. I'd love to tell you that he went so far with it that he radically found God and started coming to church. He didn't. Instead, all he did was make everyone around him miserable, most especially

22. TW: Theology has contributions to make as well. My chapter will explore the issue without committing to a "historical Adam and Eve," not because such a position isn't tenable, but because it's worth asking both whether Genesis 2–3 truly teach that, as well as whether orthodox biblical Christian faith "needs" Adam and Eve to have been two literal beings who ate literal forbidden fruit and were expelled from a literal garden. However our positions on Adam and Eve might differ, I agree that the questions Douglas poses here are legitimate and the work ahead of us necessary, important, and rewarding.

himself; as his sort-of pastor, I tried to listen and be kind, but he just could not escape "science."

In the Bible, we often read that God rejects human knowledge and understanding (e.g., Psalm 14:2–3, 49:20; Isaiah 56:11; 1 Corinthians 1–2). We are to exhibit wisdom from God not "wisdom of the world" (1 Corinthians 1:20). However, I do not believe that science itself is "wisdom of the world" in this sense.[23] Science is actually a marshaling of the creative power of humanity that demonstrates the image of God that we bear. Taken figuratively, scientists engage themselves in naming the animals (Genesis 2:19–20). "Science" undermines this, not because it is inherently wrong to interpret science, but because the theological and philosophical ideologies ("human wisdom") that come with "science" are themselves false. This is not a new development; these ideologies have existed since the beginning of human civilization. They were false then and they are false now.

When we consider the last two hundred years, science and technology have done tremendous good in our world.[24] From reducing infant mortality rate to using blood to identify illnesses to landing on the moon to developing the COVID-19 vaccines, these tangible fruits of science have blessed people. This doesn't negate human responsibility in using science and technology, nor downplay the evil that humans can do and have done through science and technology. This is where (true) wisdom and accountability come in, something that is true in all fields of study and work.

23. TW: Perhaps not. But haven't Jesus' cross and resurrection made "foolish" these kinds of human efforts to understand things, which never could have fathomed the truest nature of reality? Signs and wisdom, scribes and sages (1 Corinthians 1:20–22) are all good things, and so is conventional science, but even their best all stand forever humbled by God's surpassing "foolishness" (1:25). Theology too.

24. For further discussion, see Estes, *Braving the Future*.

It feels like you are asking me to question faith and the Bible more than you are asking me to question science and evolution. Shouldn't we question science before we question the Bible?

To be fair: Just as there is evolution and "evolution," and science and "science," so too is there the Bible and "the Bible." Our cultural understanding of the Bible shapes our reading of the Bible to some degree. For many people, especially non-Christians and "cultural Christians," but also many regular Christians, it is a rather large degree. This is why Christians need to be careful when we say "the Bible says," since we are speaking for God; therefore, we must be as accurate as possible.

We (April, Telford, and myself) are not asking you to dissect your faith or the Bible itself. We are asking you to continue to examine your interpretation of the Bible as it relates to creation and the world around us. Not all interpretations of the Bible are equally sound; and when it comes to more difficult passages like Genesis 1–2, for example, we should investigate a wide variety of interpreters. This is not to discredit traditional views per se; even among traditional interpretations of Genesis 1–2 there are stronger and weaker interpretations. Truly, we encourage you to grow deeper in your understanding of *both* science and the Bible. We should be ready to grow in both in order to be well-formed people who can navigate their culture well. We should test both our interpretations of the Bible and our knowledge about science.

As a professor, teaching students something new is easy; teaching students something they know incorrectly is very hard. These students don't need to learn, they need to unlearn, and unlearning is a daunting task. Growing up in the West, everyone learns about "the Bible" and "evolution," which creates a great deal of confusion for most people—and leads to the idea that the Bible is not compatible with evolution. It is true that "the Bible" is probably not compatible with "evolution," since these cultural ideas are intentionally set in contrast with each other. But if we are pursuing a better understanding of the Bible and evolution, I believe we will see a great deal of common ground even if we still have more questions than answers.

There are so many similarities in the way that we think about the Bible and science that we can actually use methods we use in teaching the Bible to illustrate science as well. For example, when we read the Bible, what happens is that God's perfect (infallible) word is internalized within the mind of an imperfect (fallible) person; when that person externalizes this word, it becomes an interpretation—but it is always an imperfect interpretation:

FIGURE 9

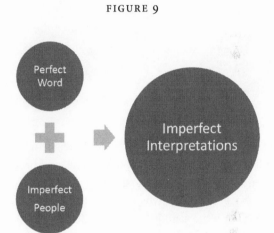

The reason is that imperfect people cannot understand perfectly God's perfect word. Imperfect people can only understand God's word imperfectly. Imperfection, though, is a matter of degree. This is why simple passages are easier to interpret, and an interpreter with skill, the Holy Spirit, experience, and more can develop strong interpretations. Yet, these interpretations are always imperfect.

Similarly, science experiences a similar phenomenon: When scientists collect data through observation and testing, assuming all goes well, then that data is accurate. But scientists have to create studies from this data, and they are imperfect. What happens next is that imperfect scientists create imperfect studies even though they base it on good data:[25]

FIGURE 10

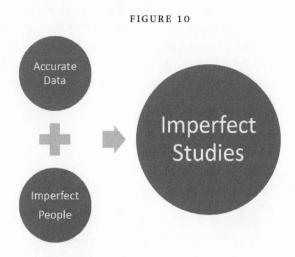

This is part of the reason science is always challenging itself, because it is always finding new and better data, and producing new and better studies. Just as above, imperfection and inaccuracy are a matter of degree. This is why if we try to interpret "Jesus wept" (John 11:35) or write a study on the Mendelian inheritance of eye color in humans, we will get a reasonably accurate answer. But as we encounter harder problems, these problems start to affect the accuracy of the study.[26]

25. Within these imperfect studies are a small percentage of actual, flawed studies. For example, a recent discussion of retracted papers reveals that around 1 percent of published scientific studies are fully flawed (some estimates show as low as 0.02 percent, some as high as 2 percent); Brainard and You, "What a Massive Database of Retracted Papers Reveals about Science Publishing's 'Death Penalty.'"

26. Anecdotally, I do meet both Christians and scientists on occasion who disagree with this basic argument; what is interesting is that their reason for rejecting the argument is philosophically pretty similar (stemming from an unwillingness to admit shades of gray), even though they come from different fields and have different views of the world.

When we combine these things, here's where we see the crux of the problem:

FIGURE 11

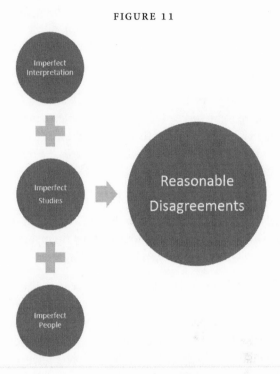

All of this contributes to reasonable disagreements—and it also leads to more questions than answers. But both theology and science are inherently question-asking enterprises: theology asks questions about God and science about the world around us.[27]

27. TW: *Questioning* is another rich and tricky term. It can connote trusting curiosity (Mark 1:27; Luke 2:46), invitation to understand (Genesis 4:6–7; Matthew 22:41), doubt (Matthew 9:14), incomprehension (Mark 9:9–10; John 16:30), skepticism (Mark 2:6–8; John 18:38), accusation (Genesis 3:11; Luke 22:23 and 23:9; Job 38:3), duplicity (Genesis 3:1), or challenge (Genesis 4:9 and 4:10; Matthew 21:23–25). Questioning can also be shut down by associating an inquiry with one type when it's really another. A question implies a relationship. How fruitful "questioning the Bible" and "questioning evolution" will turn out to be depends in part on the spirit of our questioning. That's also true of our *not* questioning them, or encouraging or discouraging others'

This leads us to one final concern: If we are to investigate and check both our ideas about what the Bible says and what evolutionary biology discovers, which one do we start with? Meaning, do we have a clear understanding of the Bible and then try to integrate evolution, or do we have a clear understanding of evolution and then try to integrate the Bible? This is an immensely philosophical question, since many books that discuss faith and evolution tend to start with one and move to the other (often, take one and fit the other into it). Personally, I believe it is a paradox, or a circle of sorts, where to be successful we must investigate the one, then the other, then back to the one, then back to the other, and around and around we go—and when we get to the stop, the truth of it all God knows.[28]

We often hear slogans like "trust the science" yet the people who argue this often do not show themselves to be trustworthy. Why should we?

Let's agree at the outset that slogans make for both bad theology and bad science. From "once saved, always saved" to "trust the science," these are examples of slogans that (while generally accurate) hide a multitude of issues. Often these slogans have more to do with "the Bible" and "science" than the Bible and science, and we should generally avoid them. Better to faithfully wrestle with the messiness that the slogan attempts to hide. To do so requires nuance, yet nuance is not a cultural value in the age in which we live.

---

questioning.

28. In the study of the Bible, we refer to this as the hermeneutical circle. TW: I agree. Both traditions need to expect to do some adapting. I'll add that adaptation will look different in each. Biblical faith is "bigger" in the sense of focusing on more ultimate matters than science. Integrating a "smaller" tradition of inquiry is a different process than integrating a "larger" one. Biblical interpretation is also historical rather than experimental in method. Our culture is actually quite experienced at this kind of integration. Christian Europe integrated Aristotelian sensibilities (with mixed results) and then Copernican astronomy more successfully, though with a few famous (and commonly misunderstood) hiccups.

But to be thoughtful Christians and reflect deeply on issues so that we may communicate with wisdom requires nuance.

Although most pastors/theologians and most scientists work with integrity in their respective disciplines, there are and always will be unreliable characters in every field of study and practice, including the Bible and evolution. For this reason, we want to be wise about how much trust we give to every scientist.[29] What makes this especially challenging today is the power that those who represent science have in our era. As I wrote in *Braving the Future*:

> Scientists have come to occupy the role of priests in our society. Scientists make pronouncements of truth as if from on high, and people are expected to follow their lead. Among the general population, most people do just that. In contrast, Christians do not see scientists as their priests (as we have a superior High Priest), and we are sometimes slower to accept new pronouncements.[30]

I wrote this before the COVID-19 pandemic, yet that period in our history demonstrated even more clearly the priestly role that scientists now play in our culture.[31] This doesn't include how culture, media, and politics warp what science says and what the Bible says.

One thing that always surprises me is the reaction to famous pastors or Christians falling away as a result of trial or temptation in their life. Inevitably, people claim they have or are walking away from the faith as a result of the actions of someone else. Both the falling away, and the reaction, are terribly sad, and we pray for returns, but I (almost) never understand why someone would quit God because of the actions of another person. Anyone who has ever engaged with the church understands that people are sinful. And it is heartbreakingly disappointing when the source of that sin are people we trust and who claim to be called by God—the priest, the pastor. Yet, they are not God, they are sinful people, and we

29. The same is true of theologians. Simply because a person claims special study of the Bible does not mean one's interpretations are solid.

30. Estes, *Braving the Future*, 17–18.

31. Harari, *Homo Deus*.

must never base our view of God on the actions and lifestyles of individual faith leaders. Likewise, the priests of the current age—scientists—are sinful people also, and we must never base our view of science on the actions or lifestyles of individual scientists.[32] Instead, we must build our faith on God and his word, and we must build our understanding of evolution on the observations and data of the field.[33] We should always go to the source to confirm what we believe.

Going to the source also helps us to not inadvertently accept and reuse weak or simplistic (straw man) arguments against (or for) evolution (or God). The media and our culture love to use these simplistic arguments, packaged as slogans, to fuel debate and conflict—which we should avoid at all costs to successfully engage our culture. Christians certainly do not have all the answers, but the answers we do have (about who Jesus is and what he has done for us) are worth defending. This should always remain our primary focus.

32. Two recent examples are Neil Ferguson, a chief scientist and epidemiologist from the UK who resigned in disgrace from his public, governmental role; and Ravi Zacharias, a Christian apologist who sexually abused multiple women. TW: Personally I want to concur by saying that I keep learning, the hard way, that Jesus reigns far above those middling and flawed powers and principalities that structure our lives. So I don't put my ultimate trust in any of them, let alone myself. "No one is good but God alone" (Luke 18:19).

33. TW: Douglas's great answer here reminds me of the ways Christians responded to the widespread and obvious failures of medieval Western Christianity: with efforts to reform rather than with wholesale rejection. The German, Swiss, English, *and* Catholic reformations (the Catholic one centering in the Council of Trent in 1545–63, but not restricted to it) all addressed the severe crisis of legitimacy that the church faced, while affirming their trust in the elements of the tradition that remained trustworthy. Addressing distortion, corruption, and failure in scientific professional circles calls not for abandoning the enterprise but for nuance—restoring and improving it by going "back to the sources," just as it does for Christian institutions. As Calvinists say, "the church reformed and always reforming."

How do we support someone who's learning about evolution?

In many ways, this is the hardest question of all. There is no easy pathway to learning evolution that will make those steeped in "evolution" and those steeped in "the Bible" happy or satisfied. It is always easier to pick a tribe and become a partisan rather than to keep learning and keep growing, to become a person of depth and nuance. Thus, the best way I believe we can support people learning evolution is to model a learner's heart, to pray for them, and to encourage them to ask their questions within faithful Christian community (you can read *church* if you'd like). These may seem rather obvious, but it's the obvious things like these that are both the easiest to do and the hardest to pull off. In whatever way it matters, April, Telford, and I all support you, if even by these words, if even from a distance.

For further reading:

Owen Gingerich, *God's Universe.*
Michael LeFebvre, *The Liturgy of Creation.*

3

A Theological Perspective

TELFORD WORK

Be straight with me.
What do you believe and why are you doing this?

FOR OLD TIME'S SAKE, I'm going to go through the five fundamentals that inspired the Christian fundamentalist movement over a hundred years ago. I believe that Jesus Christ is (1) fully divine, (2) was born of the Virgin Mary, (3) died as God's substitutionary atonement (and more) for our sins, and (4) rose bodily from the dead as our (and my own) Lord and Savior. The Bible that proclaims all this is, as my evangelical college puts it, (5) "God-breathed and true, without error in all it teaches," and "the supreme authority and only infallible guide for Christian faith and conduct." That of course includes the opening chapters of Genesis and the Bible's other creation literature. This is not to say that I'm a "fundamentalist" with all that term's popular connotations.[1] Many fundamentalists wouldn't want me. Nevertheless, I'm on board

---

1. DE: As we see in each chapter, often there are key terms that mean one thing in popular culture and mean something else in historical context or academic circles.

with all five so-called fundamentals. I'm also a special fan of the Pentecostal/charismatic tradition that expects miraculous signs of the Holy Spirit's presence, power, and fruitfulness today. I could affirm much more, and a lot of it would be considered a pretty traditional conservative picture. Not all of it, but plenty. My answers don't throw the Bible under a Darwinian bus.

And what do I believe about the origins of life? I accept the broad consensus among biologists, geologists, and physicists that the known universe is over 14 billion years old, the earth around 4.5 billion, and life on earth well over 3 billion. The findings from their various disciplines all reinforce this picture. I also accept the evidence that indicates human beings and many other primates have common prehuman ancestors, indeed that all life on earth shares common ancestors, which have changed and differentiated over the ages through a mainly (whether or not entirely) natural process of evolution. Evolution's mechanisms aren't entirely understood, but our understanding of them is growing rapidly; the exploding field of genetics is a mother lode of information. Some of the jumps along the way from the first organic molecules to us are still obscure—for instance, how cells developed, or the shift from single-celled to multicellular specialized organisms. In fact, experts still disagree about significant details of humanity's evolutionary history.[2,3] But the broad picture is solid and explains a growing variety of phenomena from across disciplines. Evolution deserves the respect it has earned as a paradigm for understanding life. This is also the dominant position, if not the unanimous one, among the science faculty, and probably the whole faculty, at the evangelical college where I teach theology. So nothing I've said so far is controversial in my community.

2. Almécija et al., "Fossil Apes and Human Evolution."

3. AC: I always feel a bit uncomfortable when I read a statement that says scientists disagree about something because many assume that means that we don't agree on anything. Scientists agree that humans evolved from mammalian ancestors that were not *Homo sapiens*; it's just that the exact lineage is still being uncovered. Scientists are always working on the periphery or details, but agree on the central tenets.

I wrote this chapter because many doubt that a reasonable Christian can think both of these things at once. I grew up thinking science discredited biblical faith, then as a new Christian wondered whether biblical faith discredited science, at least regarding human origins. It took a while to see there was much less tension between these than I had been led to believe.

If you believe in both Scripture and evolution, then how are you interpreting Genesis, and what gives you the right to read it that way?

If the Bible is our authority—and Jesus and the apostles certainly treat holy Scripture that way—then *we* don't have the right to interpret it however we like. We need to honor what it *is*. So what kind of writing is Genesis, and how do we know that?

The word *Bible* literally means "books." The Bible is a library: a list, or a bookshelf.[4] The writings on that list are Israel's and the church's highest authorities. Note that many *kinds* of writings are on that shelf: historical narratives, worship songs, proverbs, regulations, prophecies, sermons and meditations, philosophy, and even ordinary correspondence. We need to celebrate *and* honor what kind of writing each one is. You don't treat gloomy philosophy like Ecclesiastes as if it were a powerful prophecy like Daniel. And you certainly don't treat a Greco-Roman letter like Philemon as if it were an American email or a medieval European royal proclamation. Scholars, liberal and conservative alike, learn how to treat these documents by studying their historical backgrounds and literary genres. They learn to recognize the standard elements of a genre: in Philemon's case, the greeting, the summary thanksgiving, and so on.

Moreover, people switch genres all the time when we speak and write. We expect each other to catch the transitions and adjust immediately. Once upon a time, grizzly bears roamed California; now the only one left is the one on our state flag. Once upon a time

4. DE: I use the example of an anthology, in that it shows canonical intent and specific relationship.

there were three bears: a papa bear, a mama bear, and a baby bear. Those two sentences have matching forms and opening phrases, but fluent English speakers know the first sentence is a historical observation and the second one is a fairy tale. We also combine genres: Once upon a time, three bears walked into a bar. Now I've subverted the fairy tale genre by twisting it into a joke. I can get away with unannounced switches like these if my readers share my culture, but not with audiences unfamiliar with America or not fluent in English. This paragraph would hopelessly confuse some of my students, especially those I've taught abroad.

Genesis, and every other biblical writing, originated in a different world. No matter how familiar we are with the Bible, we didn't grow up with its foreign literary conventions, cultural assumptions, or lifestyles. There are ways to improve our interpretive skills, though. Scholars can study biblical languages to get a better sense of special phrases and literary devices—the Bible's versions of "once upon a time." We can read the literatures of the biblical authors' ancient neighbors and look for resemblances and borrowings. (For instance, a number of passages in Proverbs seem to originate in Egyptian proverbs.) We can set aside our expectations of how these texts *ought* to communicate, and pay special attention to the literary devices they *do* use. All this amounts to listening carefully. We're sitting at elders' feet and weighing whatever they choose to tell us. That's the opposite of pleading our right to interpret in the way we wish.

Scholars have long noted that elements in Genesis's first eleven chapters resemble similar elements in the creation myths of Israel's ancient Near Eastern neighbors.[5] However, those elements are put to a totally different use. Those nations' gods and idols are not responsible for our world! Its one source is the God of Israel. Despite all our world's sabotage and treason, he is trustworthy and tenacious in steering it mysteriously to the happy ending he intends.

5. Michael Heiser is one evangelical scholar who pursues these connections fearlessly and insists that we read Scripture in its original contexts. His websites, drmsh.com and thedivinecouncil.com, are worth a visit.

Genesis begins with two creation accounts: 1:1–2:4a, and 2:4b–25. For convenience I'll call these Genesis 1 and Genesis 2 (or Genesis 2–3 if we're including the fall and expulsion from Eden). Both accounts show us where it all starts, but they differ in stark ways, including sequence and tone. (For that matter, so does Psalm 104 with its distinct sequence and imagery of creation.) The first account puts creation's events in a famous sequence: heavens and earth, sky and sea, earth and plants, sun and moon and stars, birds and sea creatures, animals and humans, and finally a day of rest. The second account has a very different sequence: heavens and earth, first human, plants and garden, wild beasts, and finally woman from man. These chronologies don't align literally, though many have tried to force them to—out of a conviction that if they are holy Scripture they must. The contrast signals readers that the sequences mean something *other* than the kind of literal reporting we expect. If you saw a film that depicted the same events in two very different ways, with different sequences and styles and characters, you'd notice the deliberate contrast and ask, "What is the director trying to do here?"

The first sequence divides into two stages as follows, with markers bookending the sequence. Note how what God creates on day N is what God populates on day $N+3$:

	"formless"	"void" (1:2)	
1	light / darkness: day / night	"day-light" (sun) / "night-light" (moon), stars: to rule their times	4
2	water / sky	water-creatures / sky-birds: to increase	5
3	earth / seas, vegetation	earth-creatures / humans in God's image: all to increase and eat vegetation, humans to rule	6
	"finished"	"arrayed" (2:1)	
7	*shabbat* (seventh): rest and blessing		

It's an elegant, beautiful setup. What starts as unformed and empty is formed, sector by sector, then filled in the same order. The seventh day breaks from the parallelism to hallow the Jewish week, whose sabbath honors and relies on Israel's good Creator and Provider. The next story begins again with a second sequence, so why consider either order definitive and expect the sciences to confirm it? Treating Genesis 1's days as standing for ages of indeterminate length doesn't reconcile the two, or resolve conflicts with scientific pictures. Instead, read Genesis 1 as the kind(s) of literature it shows itself to be, let it say what it wants in the idioms of its original audiences, and let it use the devices of literary progression and sequence as it wills.[6]

The second account zooms in on the unique relationship between God and humanity. Its close-up shots contrast with the first one's panoramas. Strikingly, it has God making *adam* before the trees and plants, before the beasts, even before the garden. The first story is playful with Hebrew (*tohu wa bohu* in 1:2 is a well-known example); the second more so. The Hebrew word for humanity is *adam*, and Adam is made of *adamah*—the dust of the ground. The garden is set on a plain that somehow becomes the headwaters of four great rivers—something we don't see in the natural world, where streams start high in the mountains as tributaries and converge into great rivers rather than the other way around. Throughout this story there are symbolic names, puns, wordplay, and portentous objects: Adam (meaning human being), Eden (whose root word means "plain" or "steppe"), a tree of life and a tree of knowledge, naked (*'arummim*) people contrasted with a shrewd (*'arum*) talking snake (*nakash*, a word that means both ordinary slithering reptiles and intimidating dragons and sea monsters), fruit that opens eyes, God on an afternoon stroll, and the name Eve (*heva*, living one, since God announced consequences of the couple's sin that implied she would survive and give birth, not die that day). All this symbolism and figuration implies a symbolic and figural narrative. These two different accounts of creation

6. DE: Telford's reading of Genesis 1–3 is one way scholars today try to make sense of the unique styles in a text that originated thousands of years ago. Although we will never be certain, with precision, what the biblical author was trying to say exactly—in the way it would make sense for people in the author's time and place—what we can be certain of is that a simple, literal, one-dimensional reading of the text is *not* what that author was trying to say.

come from cultures and contexts and have purposes that may not line up with ours.[7]

Tim Keller, an influential Christian leader, affirms the evolution of species yet treats Adam and Eve as fairly straightforward historical characters.[8] Computational biologist S. Joshua Swamidass considers recent and universal male and female ancestors for humanity not only scientifically feasible but mathematically probable.[9] Either interpretation would make evolution and a literal reading of Genesis easier to reconcile. But Eden's scene may depict our origins less literally. Some interpreters see exiled Jews using Israel's creation traditions to counter the pagan mythology of their Babylonian captors.[10] Whether or not that's right is not my concern here. One way or another, Genesis 1–3 are inspired by God and still reveal a real past, whose promise and brokenness we still inhabit. In the words of my Old Testament colleague Tremper Longman III, they are "heavily stylized *history*."

That's no contradiction: the Bible uses symbolism to communicate all kinds of things, including history. Consider the prophet Nathan's sneaky parable that snared guilty King David (2 Samuel 12), Isaiah's parable of the vineyard (Isaiah 5), and Jesus' many parables that symbolically portrayed the very real events of his ministry (for instance, Mark 12). It wouldn't be a stretch for the Eden story to be a parable or mythologized history.

7. DE: Recently I edited a volume—*The Tree of Life*—exploring one of the symbols found in the creation story. Through this process I came to believe the biblical text intends for creation elements to be read as if there are several layers. For example, the tree of life speaks to the reader from a literal aspect, a figurative aspect, and a symbolic aspect. These aspects are not competing but complementary.

8. Keller, *Reason for God*, 97–98.

9. Swamidass, *Genealogical Adam and Eve*.

10. Regarding Genesis 1, an accessible introduction is Enns, "Genesis 1 and a Babylonian Creation Story." Regarding Genesis 2, see Rosenberg, "Biblical Narrative," 53.

But if we decide not to take Genesis literally, where do we draw the line? When does Genesis start being true? Doesn't affirming evolution turn the Bible into merely a collection of human opinions that we can't rely on?

It's not up to *us* to draw those lines. If you tell me about the California state flag but start with "once upon a time," it's irresponsible for me to turn what you said into a fairy tale. The question is whether the clues in Genesis 2–3 guide us to treat it as a figural narrative about human origins or a literal one. That is a totally different question from whether Genesis 2–3 is *true* or even inspired. When Jesus tells a parable, he is being figural. You might even say he's crafting "fiction," though that's not really the right word. Yet his parables are also inspired, true, reliable, and authoritative.

It's true that some people "choose" to treat the whole Bible as fiction. But that doesn't *make* it fiction. It just shows ignorance, poor judgment, oversimplification, or wishful thinking. Others claim the Bible is mythology, despite how much obviously isn't. And some choose to treat the whole Bible as literal nonfiction, despite the many signs that some of it doesn't fit that category either. The Bible is a whole collection of writings that are doing different things. God's word quite obviously taps into the full richness of human communication. It's one reason the Bible has captured imaginations for thousands of years.

Often a text is doing more than one thing at a time, and maybe even more than the author intended. Readers *can* draw conclusions about, for instance, whether a historical narrative also conveys a moral, or perhaps takes on an unintended ironic or symbolic significance. Maybe my California flag story pierces your environmentalist conscience. The Gospels often relate irony: for instance, Caiaphas advised the Sanhedrin that it was expedient for Jesus to "die for the people" (John 11:45–53 ESV). They also find prophetic foreshadowing in some Old Testament texts that obviously weren't intended as prophecies of the Messiah: for instance, Hosea 11 in Matthew 2:15 and, more controversially, Isaiah 7:5–16 in Matthew 1:22–23. Douglas showed how readers with inadequate judgment arrive at imperfect interpretations. Picking up on multiple senses is a matter of using good judgment, not coloring outside the lines. Taking everything literally doesn't prevent imperfect interpretations; in fact, it spawns some.

If forced to choose, I'd prefer the "conservative" rule of thumb of treating the Bible literally to "liberal" rules of thumb that treat it mythologically or as limited and fallible human opinions. While these rules of thumb are still lines drawn in wrong places, Christianity is still at its core a historical tradition, about a series of real events involving Israel, Jesus, and his disciples. So literalism is a less catastrophic mistake than anti-literalism.[11] Yet evolution is an area where the cost of reflexive literalism outweighs the benefit. Not only does it keep us from reading Genesis as perceptively as we could by forcing it to act like a scientific chronology, but it robs the gospel of genuine credibility and forces people to choose between "science" and "faith" when the two don't actually conflict. In our technological era, it's creationism that looks like unreliable human opinion, and biology that looks like dependable and fruitful knowledge. Scientific narratives have greater cultural credibility nowadays, and after decades of compulsory education they come easier to us. The pressure we feel to prefer scientific to biblical depictions of origins or to "scientize" Genesis in a creationist fashion owes at least in part to greater confidence in our own culture than the Bible's stranger voices. We naturally want Scripture our way.

But we don't get it our way, and that's a blessing. Readers are responsible for familiarizing ourselves with the Bible's many backgrounds and textures. Happily, this relieves much of the pressure we feel somehow to reconcile these creation accounts to scientific findings. The Bible consistently shows us a relational God who is far more interested in our relationship than tutoring us in the mechanics of creation. Think of God's words to Job (38–39)! Of course I'm curious about those mechanics, but I'm grateful that the creation stories focus on saving us rather than catering to our curiosity. They focus on human beings' unique relationships with God, one another, the rest of creation, and ourselves. The rest of

11. DE: Fully agree.

the primordial narratives concern how sin, depravity, and judgment ruin those relationships. Together they set the stage perfectly for the "reset" through Abraham that comes in Genesis 12. That reset spawns a fellowship of saints whose life carries through to the rest of Scripture, the present day, and the end of the age. We're better off with the Bible we have than whatever Bibles we want.

The New Testament treats Adam, Eve, Abraham, Moses, David, and others in ways that don't seem figurative. They even mention biographical details. But you treat Adam and Eve as characters in something like a parable. Wouldn't people like Jesus and Paul be in error to refer to them as literal historical figures?

Genealogies[12] do link those characters in a common past that leads into the authors' present. The Bible doesn't treat Adam and Eve as unhistorical myth. That doesn't mean they are straightforward characters, though. Esteemed evangelical Old Testament scholars Tremper Longman and John Walton suggest that we talk about events in ways that span a spectrum between "metaphysical" (deep meanings) and "empirical" (surface appearances).[13] I'll use a co-ordinate plane, and illustrate the range of possibilities with some flippant contemporary examples:

12. The *toledot* formula ("these are the generations") found in Genesis 2:4 repeats later in the primeval narratives (5:1, 6:9, 11:10, 11:27), and recurs in the patriarchal narratives (Genesis 25 and 36–37).

13. Longman and Walton, *The Lost World of the Flood*, 19.

FIGURE 12

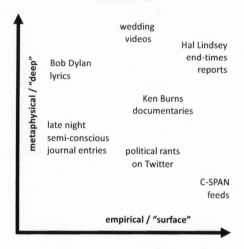

The ways we depict events clearly span both ranges. Now let's compare a few passages where Mark's Gospel describes past events (1:1) and real characters:

FIGURE 13

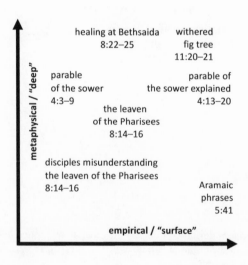

Sometimes Mark dwells on apparently trivial historical details like Jesus' exact words in Aramaic; other times he passes on parables and anecdotes laden with hidden significance. Mark's characters are often realistic, but they can also be mysterious. A nameless young man in linen appears at Jesus' arrest, and there's a young man in linen at Jesus' empty tomb (14:51–52; 16:5). Scholars debate: What do these cameos mean? Is this one man or two? Man or angel? A mere literary device? Is the man the demonized Gentile Jesus delivered in Mark 5, or the rich man who turned away in Mark 10? Does Mark want readers to connect those dots? Who knows where on that graph he, or they, belong! And sometimes Mark's purpose seems to shape his depictions: he likes to sandwich one event between two halves of a different event (see 5:21–43 among many others). Matthew and Luke take apart many of Mark's sandwiches when they describe the same events. Mark's whole Gospel also falls into three acts over two halves, so his sandwich technique may be a guide for interpreting the whole thing. Mark is certainly referring to the past, but even he depicts it in sometimes tricky ways.

Turning from one extreme to another, let's examine a New Testament treatment of Noah's great-grandfather Enoch. In Genesis 5, Enoch's character is just a passing figure in Israel's family tree. Yet his depiction resembles Enmeduranki, the seventh king in the Sumerian King List; it follows Genesis 1–11's pattern of subverting the primordial Mesopotamian sages that Israel's rival Babylonian culture worshipped. So early readers knew there's more to Enoch than meets the eye.

The comment in 5:24 that "he walked with God and he was not; for God took him" gave rise to all sorts of legends. Literature that claims to come from "Enoch" was being written until at least the third century AD. I like to call it "fan fiction." Jude mentions him by drawing on 1 Enoch, an extravagant writing from outside the biblical canon that was popular at the time. (He does something similar with the noncanonical Assumption of Moses in verse 9.) Jude quotes a section of 1 Enoch dating from about 170 BC, as if its words came right from Enoch.

Could Enoch have said all that Jude says he said? Sure. For all we know Enoch said America's Pledge of Allegiance. But no reasonable historian thinks of anything in 1 Enoch as the patriarch's actual words. They are at best legendary—at zero on my scale's "empirical" axis. Yet Jude repeats them alongside other legends that were current in his day.

Why can't we get away with this today? We do, all the time. Think about how I might turn one of Clint Eastwood's iconic characters into a sermon illustration: "As Harry Callahan said to a criminal as he trained his gun on him, 'Go ahead . . . make my day.'" I might even describe Callahan as a middle-aged, frustrated Californian who grimaces whenever he sees injustice being done. I don't need to tell anyone that (Dirty) Harry is actually a fictional character, because even long after the films most of you still know that. Fifty more years from now, people reading my sermon might not get the reference. Would that make my sermon false or inaccurate, or mean I was misleading my congregation, or even mistaken myself? No. Would that future audience be misled? Not really, because they didn't belong to the world I shared with my original audience.

Yes, but Dirty Harry really *is* fictional, you say. Jude would have been ridiculed for appealing to what is depicted as a fictional character in Matthew 22 and Acts 2! How can Dirty Harry compare to Enoch or Moses? Well then, let me ask you to imagine that I'm quoting Oliver Stone's *JFK* instead. I might say, "As John F. Kennedy said," and quote some of the screenplay. I might mention the dates of JFK's birth, presidential term, and death. Now you might be a historian of JFK, and you might know how much Oliver Stone likes to put words in the mouths of his characters in his (misleading) docudramas, and so you might tell me politely after the sermon that those weren't JFK's real words. Oh, I might reply; well, no matter, the line still makes my point. Both the movie character and the real guy are named John F. Kennedy. I didn't make it clear which one I was quoting. After all, I wasn't delivering a historical lecture on the man or endorsing Stone's conspiracy theory; I was using him to teach something else. The teaching stands. Note that

I'm not even treating Dirty Harry or JFK as figural here; in each case I'm quoting a literal character in a story to illustrate my point.

What if Jude and other biblical writers were doing that? When they appeal to Adam and Eve, Enoch, Moses, Abraham, David, and the like, they aren't quoting flesh-and-blood people with whom they've had personal conversations. They're referring to characters in a familiar narrative—larger-than-life characters, in fact. Each one is not only a man, but *also* a legend, like JFK.

How then can we say Jude is inspired, if Enoch might not have said what Jude says he said? Well, let me rephrase that question: "How *does* biblical inspiration work if it allows Jude to appeal to Enoch the fan fic legend rather than just Genesis's cameo version?" Rather than assuming that the Bible is illegitimate if it does things that sometimes fall short of *modern* standards of historical accuracy, we should ask how God's word *is* using these sources, and let the *actual evidence* shape our understanding of inspiration, rather than letting our preconceptions about inspiration drive the way we "allow" the New Testament to use sources in and beyond the Old Testament.

Let's draw another spectrum:

FIGURE 14

"the man"   "the legend"

I'd put King David and JFK toward its left side, Enoch toward the right, Dirty Harry on the far right with the Bible's very few clearly fictional biblical characters, and Moses somewhere in the middle. Lots of the Bible's mundane historical figures—Rufus and his mother, for instance (Romans 16:13)—go firmly on the left: all man and woman, no legend. And what about Adam and Eve? I'll put them in the middle, to the right of David and Moses, with an asterisk. In Genesis 2–3 they symbolize our ancestors who were first called by God to discipleship and abused their power. There's no reason to treat those ancestors as any less real than JFK. Adam and Eve are no *mere* myths or legends; they vividly depict whatever real flesh-and-blood ancestors kicked off humanity's theological drama. They depict the human beings who originally betrayed the God who had called them into fellowship and who left us all mired in their unpayable debt until Christ could come and free us. But that inevitably makes them legends; they depict that tragic kickoff *as* symbols that take on mythological significance. Their characters have expanded wildly in the lore of Jews, Christians, Muslims, and all humanity. They're archetypes too, since each new generation of sinners immediately sees ourselves in them. No wonder Paul sees his whole dysfunctional congregation in Corinth in Eve's deception and disobedience (2 Corinthians 11:1–5). I doubt we'd find the real-life persons in a Christian paleontologist's realistic documentary so alluring and convicting.[14]

I don't see evolution posing a new problem here. Paul's analogies in Romans 5 and 1 Corinthians 15 don't depend on Adam lacking *prehuman* ancestors. They don't even fail if he's a figural character. Genesis 2–3 depict aspects of our past that Jesus came and died to address. Sin began in some concrete way after the beginning of human history, and it has cascaded down that history to afflict us all. The concrete death of Jesus reverses its effects in ways that pertain to us all, cascading similarly to us as we trust the good news. Jesus' resurrection has inaugurated a new "origin story" that fulfills the old one. Paul says that in him our old-Adamic "soulish bodies" ("soulish" reflects the Greek word) will be transformed into glorious imperishable "spiritish bodies" like his (1 Corinthians 15:42–53). We've borne the image of that original human of

14. DE: Telford and I thought it would be helpful to mention my view of Adam and Eve. One challenge, though, is the space to develop what is a complex discussion for all scholars. The best simple way I can explain my view is that I work the problem somewhat differently than Telford; I come to a slightly different conclusion at each step. As a result, for me Adam and Eve end up further left on Telford's chart than he places them. In saying this, I strongly urge readers to remember this is a complicated issue and not to draw black-and-white conclusions from this. I agree with Telford (really, historians everywhere) that all historical figures fit on this spectrum somewhere, and that all have legendary features (which seemingly is why we have the discipline of "history" to begin with).

dust, but we'll bear the image of the original human of heaven. Paul's analysis of Jesus' atonement works whether or not "Adam" refers to one and only one flesh-and-blood human being without a belly button.[15]

Does evolution force me to give up the belief in Psalm 139 that you and I are fearfully and wonderfully made just as God intended?

Don't give up Psalm 139! Christians who embrace evolution still pray that psalm along with the rest, for the same reasons as creationists. After all, when you and I pray "you knitted me together in my mother's womb" (Psalm 139:13 ESV), we aren't denying the natural reproductive processes behind our personal lives, nor are we praying to a "God of the gaps" who mysteriously does only the things we can't explain as nature. We're crediting God for all those processes, whether or not we understand them. I know the processes by which my children came into being, but I don't hesitate to say that God gave them to us, and knows and loves them personally. Physicians and biologists know those processes a whole lot better, but they don't hesitate either. This isn't new. Augustine's and John Chrysostom's commentaries on Psalm 139 don't speculate about what God might actually be doing in nurturing a fetus.[16]

We grow food in sun, soil, and water; we ship it, shop for it, cook it, and serve it; and then before we eat we thank God for it, despite all the work we obviously did ourselves. Saying grace is not some holdover from premodern times; if anything, the ancients knew how food grew better than most of us today. Mealtime blessings respect that creation's natural processes owe somehow

15. DE: This may be true, although I personally wouldn't arrive at this conclusion. I think it is the very reality of Adam's body that makes the analogy work, and work so well.

16. Augustine's *Exposition on Psalm 139* makes little distinction between human gestation and any other: "*You have fashioned me, and hast laid Your hand upon me.* Fashioned me, where? In this mortality; now, to the toils whereunto we all are born. For none is born, but God has fashioned him in his mother's womb; nor is there any creature, whereof God is not the Fashioner."

to God's good agency, providence, and intentions. Embracing evolution simply means extending this respect for God's agency, providence, and intentions for individual organisms and prepared meals to whole species. You are still fearfully and wonderfully made. So are all species under the sun, whether they owe to natural processes, supernatural ones—or even artificial ones. (Did you know that oranges are hybrids? Our ancestors combined pomelos and mandarins, but they are still wonderfully made.)

Evolution does, however, influence how we *imagine* God making earthly life. Evolution evokes the wonder and fear we feel for wilderness rather than a curated garden. The Bible has room for that, too. Creation looks a lot more intimidating, rough, and less hospitable to us humans in Psalm 104 and Job 38–39 than in harmonious and orderly Genesis, and God's creating looks a lot more hands-off in Genesis 1's "let there be" than Genesis 2's "God formed." The wild, hands-off picture has grown on me. Evolution eases the challenge that nature's terrors pose to God's goodness: predation (Psalm 104:21) and parasitism, life cycles (Psalm 104:27–29), mutations and birth defects (Exodus 4:11; John 9:3), environmental catastrophes, food chains, and extinctions. Paul says we endure these things while we wait for their Creator to release creation from its present subjugation to transience and decay (Romans 8:21–22) and make all things new. An exclusively hands-on picture of God's creative work implies that all these features were designed in—including every nuisance from mosquitoes to tarantula hawks, and even viruses—so we'd better be grateful and grow out of our squeamishness. That is how nineteenth-century natural theology framed the natural world in Charles Darwin's day, and he and his fellow Victorians rightfully found it discouraging and fatal to their faith as they uncovered more and more unpleasantness. By contrast, evolution's hands-off picture suggests that these unhappy sides of nature are possibly necessary features of a cosmic order that, while largely inhospitable, supports terrestrial life while humanity walks Christ's way of glorification. Evolution also helps us see the story of earthly life as one of chronic change—which the evidence demonstrates anyway—rather than some originally pristine Edenic perfection. Since the perfection we await is resurrection to *new* creation, evolution helps us resist the cultural habit of looking backward for a "paradise lost" that distracts us from pressing on to the better goal of resurrection from the dead in Jesus (Philippians 3:10–14). On that day, we really will be fearfully and wonderfully (re)made.

How can God and creation both be good if a lot of bad human behaviors are evolved rather than sinful?

Both God and creation can be good and are, but we have to be careful with the term *good*. To give you a full answer, I want to imagine our creation, fall, and redemption in a way that tries to be faithful to Scripture and mindful of evolutionary biology.

"God saw everything he had made, and behold, it was very good" (Genesis 1:31 ESV). Yet evolution traces thorny human traits and behaviors to our evolutionary inheritance: aggression and domination, deception, tribalism, mating preferences such as female hypergamy versus male preference for "erotic capital," adolescent recklessness . . . we could go on and on. How could these be good, when so many are sins for us?[17]

Classical Christian doctrine blamed those shortcomings on the fall of Adam and Eve. Bad animal behavior was shrugged off as raw beastliness, or else blamed on the fall too. Our usual answers assume things that aren't actually in the text. We often call Eden "perfect" and imagine salvation as getting back to that original flawless paradise, but that way of thinking is closer to Plato than Moses or Jesus. Rather than a story of *perfection* and *sin,* let me propose a story of *goodness, newness, oldness,* and *sin* where those evolved traits weren't products of the fall, nor were they meant to characterize humanity or even creation forever. They're temporary

17. AC: Evolution is not all "red in tooth and claw." The literature base in evolutionary biology and altruism is immense and the past several years has seen a new resurgence of articles and books, especially in cultural evolution. The findings suggest caring for others evolved, and we can see evidence in reptiles and mammals.

behaviors that creation is meant to transcend. Natural selection explains the value and persistence of these awkward evolved traits across species, including humanity. It *is* raw beastliness—and that's okay. *Good* even.

The Hebrew *tov*, "good," means agreeable, appealing, excellent, fruitful, valuable, appropriate, and upright. Its range of meaning isn't so different from our English word. Think of all the ways a wristwatch or pair of pliers could be good. Just don't use a watch as pliers. Our kids grew up reading Eric Carle's book *The Very Hungry Caterpillar*. Caterpillars are good. So are butterflies. But good caterpillar behavior is terrible butterfly behavior, and vice versa.

Let's put Genesis 1:31 this way: Creation was an altogether good start. Then Genesis 2:5 and 2:15 insist that human labor was necessary to cultivate the earth as God intended. So Genesis's creation accounts have the feel of preparations for a big event. Our divine host has readied his venue and called his workers to *begin* helping actualize its potential. Creation was a work in progress even before sin ruined things. The rest of the Bible, beginning with the second creation account, reveals what was to come next: pupation to butterfly life—being transformed into *new creation* and united through fellowship with its transcendent and utterly glorious Creator (Isaiah 25:6–8, 65:17–25, 66:22–23). That's better than good! It's perfect.

When I was a new Christian, the picture of a perfect paradise lost loomed so large in my imagination that I couldn't see the biblical details that didn't fit. But evolution posed such a formidable challenge to it that I had to take a fresh look at the Scriptures. I was surprised to find *more* room for biblical Christianity than before. Especially in Paul: Original creation isn't fit for that future without a resurrection to a whole new state of being: "Flesh and blood cannot inherit the kingdom of God, nor does the perishable inherit the imperishable" (1 Corinthians 15:50 ESV). Mere mortal flesh and blood suffers and dies, to rise as God's eternal heirs (Romans 8:16–17).

Perfection demands new ways of life for which some of nature's old ways are unsuited (15:39–49). Evolved traits like empathy and courage belong in both stages, but aggression, selfishness, and undisciplined emotion don't cross over.

Biology provides an obvious analogy: adulthood, when childhood ends (Galatians 3:23–4:9). Our Darwinian past belongs to creation's childhood, its caterpillar stage. "When the perfect comes, the partial will come to an end. . . . When I became a man, I ended the things of early childhood" (1 Corinthians 13:10–11).

If the survival value of clunky inherited caterpillar behaviors like selfishness and tribalism drove "old" life's origins across the animal and plant kingdoms, then they were and are good. The kingdom is even better (Matthew 11:11), calling us out of some of these and into the beautiful fresh butterfly ethics of Christ's "new" eternal life: for instance, laying down one's life for slaves and loving enemies sacrificially, which is "perfect as your heavenly Father is perfect" (Matthew 5:43–48 ESV). For the Lord behind both creation's childhood *and* adulthood is surpassingly good.

What has never been good is refusing the Lord's call out of our old life. Let's focus on mating and reproduction, which are both the heart of evolutionary biology and prominent among the New Testament's sometimes puzzling and irksome rules for Christians. Taking a mate—or more than one—makes Darwinian sense. Leaving one can too. We see this across the animal kingdom, whose mating behaviors are "good" the way wristwatches, pliers, and caterpillars are. Yet Jesus (Matthew 19:3–11, 22:30) and Paul (1 Corinthians 7) condone and even encourage celibacy, command chastity, monogamy, and sexual fidelity, and reject divorce except under narrow circumstances. At the same time, they command loyalty to Christ above tribe and family, a person's strongest allies and often one's only truly dependable support network. Putting love of enemies (Matthew 5:43–48) before familial protection (Matthew 10:34–39)? *Are you kidding?* Christian discipleship seems like applying for a Darwin Award. Why this risky and apparently inconsistent combination of putting spouse before self and fellowship before family? These rules embody hopes that old creation is ending when Christ returns and the dead rise to a new, post-Darwinian kind of life when marriage's time has passed.

Following the Lord calls us to some new behaviors that anticipate that coming future (Colossians 3), and away from some natural behaviors that interfere. Dwellers in Christ's chrysalis don't stay caterpillars (1 John 2:28–3:3).

In the Sermon on the Mount, Jesus renews and radicalizes Eden's call for us to forsake our passing lives and seek God's eternal kingdom and righteousness (Matthew 6:33). When our world and we ourselves are still childish, "adulting" is a real challenge. In mating, commerce, order, and ordinary relations we seek status, advantage, and defense for ourselves and our group. That caterpillar stuff will no longer do; his kingdom has altogether different priorities (Matthew 19:10–12; 6:19–21; 20:25–28; 6:22–23; 5:38–42). Jesus' teaching, example, and work all point his disciples beyond *both* good-but-old present conditions *and* bad-and-cursed fallen ones, and urge us to prioritize the lasting, gloriously good future he secured.

Wholeheartedly seeking the kingdom and its beatific righteousness is a big ask, and thoughtful people have balked at it from the beginning. Jesus knew that sacrificing good evolutionary advantages such as family and tribe could be a death sentence (Matthew 10:37–38). Holding onto life calls us beyond biology (10:39). Many see the kingdom as otherworldly, a disembodied life "up" in heaven sometime in the future. It's better to say that the kingdom is creation's non-Darwinian butterfly future invading the Darwinian *and* fallen present age and inviting us into the chrysalis of Jesus' body.

Sin spurns that inheritance (1 Corinthians 6:9–10). Genesis 2–3 display the first humans trying to fly, so to speak, without growing wings. Does Genesis 3:6, when Eve and Adam "took of the fruit," allude symbolically to humanity's decisive first rejection of God's way to God's future? It wouldn't surprise me at all. Isn't it just like an ape to grab that fruit? But God needed representatives, not more aping of the old order.[18]

18. AC: I know Telford read my chapter and realizes we didn't evolve from apes, but he can't resist monkeying around.

Since Jesus is a new Adam (Romans 5:15), his situation resembles that of the first human beings. God brought *adam* into a covenantal relationship. Its distant goal was Christlikeness. Given evolutionary human ancestry, taking that pilgrimage meant leaving behind some good and previously helpful evolved behaviors that might not have been sinful before, but had no place on the journey. A primate's greed is no more sinful than our family dog's, but it doesn't image God as we're intended to, so it has to stop. Maybe—I'm speculating here—God's first command was really for Adam to set aside for once his impulsive hominid curiosity and self-sufficiency and *trust God*. Faith would have taught him a whole new way to be himself.

Some struggle to reconcile human evolutionary origins with a fall from grace. But Genesis 2–3's symbolism stresses relationships forged and ruined, and an evolutionary history doesn't threaten that. The fall in Eden was the first of its kind: a catastrophic human failure to walk forward with the Lord who had called humans into newness of life. If Eden was a workplace ready for the workers to go on duty, the fall was when they disobeyed their owner, tore up the plan as it had only just begun, and enslaved all who followed to their immaturity. Whatever its literal particulars might have been, it left our ancestors worse off than the naïve beasts from which we came. Genesis 4–11 trace the spreading disorder of caterpillars indulging their insatiable appetites when it's no longer appropriate, taking flights of fancy, and causing all kinds of new problems. These are sin's consequences. Sin worsens biological death by turning it into a just consequence and punishment, a fearsome master, and an unavoidable enemy—*relational* and *spiritual* death as well as physical.

God graciously offered a new start, telling Abram to "go from your country and your kindred and your father's house" (Genesis 12:1 ESV)—to sacrifice his old environment and become a blessing to all the earth's families (Genesis 12:1–3). Genesis 12 onward remember God addressing the mess with a new effort that culminates in the coming of a new arrangement in the Lord Jesus, who told a scribe to "follow me, and leave the dead to bury their own dead" (Matthew 8:22 ESV). His chrysalis is our only way forward. He's the new Adam that delivers us from the old one's ugly legacy (Romans 5:12–21). In him we can not only escape the deadly consequences of past failures but inherit the transformation intended for us all along (2 Peter 1:3–4; Hebrews 13:14; Matthew 19:29).

My ambitious interpretation begs all sorts of valid questions. Of course I'm speculating in interpreting Genesis 2–3 this way in light of the rest of Genesis, creation's and God's goodness, the work of Christ, and the scientific evidence. Others reconcile them differently. What's important is that they *can* be reconciled. A solution may even improve on earlier interpretations.

How then are human beings uniquely made in God's image? Are souls part of evolution?

Christian thinkers have answered the first question in a variety of ways: We're rational, imaging our all-knowing God. We communicate with language, imaging God who spoke all things into being through his word. We relate to one another as fellow persons, imaging our one God in three persons. God gave us a task furthering his work as Creator. Our love images God who is love. God is spirit, so we have souls.

Some of these answers don't distinguish us clearly enough from other creatures. Angels, dolphins, crows, octopi, primates, and AI machines are intelligent, relational, and communicative too. These answers also raise disturbing questions such as whether certain human beings—the unborn, the developmentally disabled, the senile, even the dead—image God less than typical humans, and whether people with superior traits image God more. Evolution sharpens these old difficulties: Would prehuman ancestors have more nearly imaged God as they approached humanness? Can posthuman descendants evolve *out* of imaging God?

Seeing our *souls* as imaging God seems promising. An easy way to situate evolution happily alongside Christianity is to suppose that *Homo sapiens* became both "human" and *imago dei* when God bestowed immortal souls on them. It's hard to see this as an evolutionary biological process. But either way we run into some wrinkles. What about angels? They have some kind of eternal immaterial existence that's compatible with fellowship with God. How different are they from how Christians typically think of human souls?[19] And sentimental Americans seem convinced that their family pets have souls and afterlives. There's no biblical support for that, but it shows that we use the concept of soul in a way that doesn't pinpoint how humans uniquely image God.

A better answer for how we uniquely image God is *vocation*. The specifically human task of being God's representatives as rulers and stewards—living agents and subjects of God's providential rule—is delegated only to human beings (Genesis 1:26–27, 9:6). Our unique vocation doesn't depend on dissimilarity from other kinds of organisms or angelic creatures.[20]

God has graciously granted some of his creatures a certain appointment meant to culminate in our adoption as "sons" or heirs in Christ—when we can call Jesus brother, friend, and Lord, name the one who sent him Abba, receive his Spirit to dwell in us as a temple, and partake in the divine nature (2 Peter 1:4).[21] Why regard gaining the image of God and all its responsibilities and consequences as anything *other than* receiving that relationship? We don't need to assume it involves some physical or metaphysical change, and we need not rule out one or another physical or metaphysical past. The relationship itself is decisive. That's certainly how we treat belonging in Christ: a person's new birth from above is a transformed relationship, received in faith (John 3:3–16). Evolutionary biology offers possible explanations for the rise of some human traits like intelligence that feature in our relationships with

19. Acts 12:15 shows that first-century Jews sometimes called a ghost a person's "angel."

20. April coauthored a book with Michael Lodahl on this topic, *Renewal in Love*.

21. AC: I would add that this relationship Telford describes is what makes us special, as I attempted to explain in my chapter when discussing human evolution: being appointed (Telford's term) as beloved representatives instead of being instantaneously created from dust.

God. But our ancestors didn't evolve themselves into *imago dei* relationships, nor will mutations evolve us out of them.

So far we've mainly talked about the Bible. But what does "theistic evolution" say about God's character as Creator? Is God okay with all those millions of years of death? Isn't he callous or even cruel rather than loving? The God of evolution sounds like a gambler rather than a planner, watching eons of purposeless random change rather than working to produce something intentionally good, beautiful, and sacred.

There is real force to this objection. Problems with evolution go deeper than just how to read Genesis. Evolution reshapes how we understand God's goodness. Traditional doctrines of creation portray a beautiful, harmonious original order—*Edenic!*—that is only marred once the first humans disobey their loving Creator and usher in sin, disorder, and death. That's not what paleobiology sees, to put it mildly. Just watch a few nature documentaries such as *March of the Penguins* and witness how gargantuan a struggle life is for many plant and animal species, and how finely adapted they are to it. Millions of years of subsistence living and mortality are hard to pin on Adam and Eve, not to mention meteorites, volcanoes, earthquakes, and tsunamis.

Some will blame them on an angelic fall billions of years ago. Not only is this a distorted reading of the usual Bible passages it relies on,[22] but it strains Genesis 1:31–2:3's affirmation that God's human-populated heavens and earth were "very good." Scripture would have us think of these terrifying phenomena, as well as randomness (Ecclesiastes 9:11), somehow as features of God's

22. Revelation 12 speaks of Satan being expelled from heaven not in the distant past, but by the blood of Jesus and the testimony of his martyrs in the church (12:11). Isaiah 14's account of the proud fall of the Morning Star is directed at the king of Babylon, not an angel (14:4). The immensely influential interpretation of these texts as referring to the fall of an angel long before humans were created is not well supported by the texts themselves.

good creation. Both are thorny whether God was more hands-on in shaping creation (guiding the process somehow, or even intervening supernaturally as an "intelligent designer" to force changes that wouldn't happen otherwise) or more hands-off (creating the whole system and foreseeing its results, but not interfering in its dynamics much if at all).

It's worth pausing over that issue, though. Is God more like a helicopter parent, or a free-range parent? We see God in both modes in Scripture: delivering the Hebrews from Pharaoh and micromanaging them in the wilderness, for instance, after four hundred years of just letting them multiply under oppression. The combination shapes Israel's hope for the future: the prophets assure doubters that God will act decisively to rescue his people in a way he hasn't yet. Even after all of Jesus' redeeming work, Christian hope takes the same form. It shouldn't surprise us to see either mode in prehuman history. Hebrews 11–12 remind us that God is more patient than we are *and* more determined to see things through. We've endured thousands of years of delay between the fall, God's promises to Abraham, Moses, David, and the rest, and even Christ's ministry and the present day. Evolution in an old universe doesn't really add anything truly new, just more zeroes.

Randomness pertains to God's free-range mode. Some Christian theologies make little or no provision for it, affirming instead that God intends all things (except, somehow, sin) specifically to happen. Sin must have a cause besides direct divine action in order to be sin. But quantum events do seem genuinely random, and laws of probability suggest that God allows some events rather than specifically determining everything. If I came about because one sperm happened to outswim billions of others, well, that's obviously something God can work with. If a stray cosmic ray mutates a gene and I get terminal cancer, so be it; that's life in old creation. My father died young from a fluky knee injury, changing our family's lives forever. I don't need to look for special reasons for these things or even affirm that they're somehow details of God's eternal plan. They might just be part of that pointlessness that creation can't wait to escape (Romans 8:20a), yet God can use in hope (8:20b) to bring about its fulfillment (8:21).[23]

So let's approach that emotionally powerful objection this way: How do you interpret generations of Hebrews suffering and dying in slavery, centuries of Israel's and Judah's apostasy and exile, and millennia of church persecution, confusion, corruption, and division? We're tempted to blame either God or human free will. The more biblical response is quite different: to lament evil and suffering, reflect on God's past mercies and righteousness, and hope for God to end all miseries someday.

It's no good to deny how disturbing and distressing our past and present conditions have been, or lament only our own species' predicament. In our Father's world, the birds of the sky receive their food (Matthew 6:26), but they also struggle, starve, and get preyed upon. Yet every sparrow's life is still sacred to him (Matthew 10:29). God grieves over the very death he presides over. All creation "groans in labor pains" while she waits and waits for "freedom from slavery to decay" (Romans 8:21–22). "Blessed are those who mourn, for they shall be comforted" (Matthew 5:4 ESV).

God sprinkles into history signs that death doesn't get the last word (Revelation 20:14; 1 Corinthians 15:55). They multiply to this day. These signs aren't random! They aren't signs of callousness, let alone cruelty. I tell distressed students that the lavish future God has prepared for us (1 Corinthians 2:9–10) proves he is not okay with a friend's affliction or a relative's death. God takes no joy in suffering, not even of the wicked (Ezekiel 18:23). The Lord wept outside Lazarus's tomb, even though he knew his dead friend would soon rise (John 11:35). Even if God is rather hands-off, leaving a lot to creation's internal dynamics rather than controlling it as a hand flexes a glove, that's neither gambling nor purposelessness.

23. Douglas's chapter suggests that he may have more of a hands-on vision of God's relationship with creation. Christian theology spans a spectrum of mediating positions, preferring these to the extremes of hands-on exhaustive determinism on the one hand and hands-off deism on the other.

Fullness with God (Ephesians 3:19) is how creation is finally good, beautiful, and sacred, in ways that engulf its earlier incapacities and sufferings, evolutionary or otherwise. Paul sees God's interventions moving creation from bondage toward *increasing* freedom—*and* from distance into fellowship. This goes way beyond a restoration of whatever primitive order Genesis 2 represents. It's common to plot creation, the fall, and redemption in a kind of v-shape: down, then back up. But salvation is intimate, Spirit-endowed eternal sonship, not a mere recovery operation to bring us back to the starting line! Life is sacred beyond our imagination. So let's substitute a root symbol √ that respects both the deep loss of sin and the much greater gain of new creation. Think of the starting point at the middle of the vertical as a relationship of difference, the bottom as estrangement, and the top as our ultimate likeness to God.

FIGURE 15

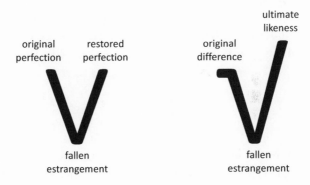

An evolutionary account fits this pattern of deepening relationship really well. Life's prehistoric evolution—that little flat part of the square root sign that actually took millions of years—brings neither alienation nor intimacy. In Genesis 1's terms, day five is more *complex* than day one, and closer to being *ready* for God's invitation to fellowship, but it's not better or more godly in itself.

That growing readiness is significant, though. As life moves up the scale of complexity, traits emerge that seem adaptive to God's goodness. Our mirror neurons facilitate empathy that moves love beyond advantageous social attachments. Hybrid vigor attracts us to outsiders and extends love, at real risk, beyond our own tribe. Intelligent organisms don't just perceive opportunities to survive and flourish but discern deep patterns, transcendent truths, even beauty. If such emergent traits can be explained by Darwinian logic, then we may be finding traces of God's goodness and purpose latent in his creation. If they cannot (altruism is tough to explain) then such "gaps" might be signs that point us beyond creation to its purposeful Creator. Both sets, not just the apparent gaps, are consistent with Israel's and the church's faith and Scriptures.

You may think evolution and Christian faith go hand in hand, but my secular friends say evolution demonstrates that science is superior to religion. Isn't evolution just a part of the whole package of secularism—a Trojan horse for a secular worldview? Jesus said we know a tree by its fruits, and Darwinism is behind so many evils of the modern age, from eugenics to national socialism to predatory capitalism. How can something like that be compatible with faith in Jesus Christ?

Why do we think faith and science cut against each other? Douglas's chapter shows that they really don't. Many of the scientists leading the scientific revolution in Europe were believers of some sort or other. Their confidence that God was trustworthy led them to view his created universe as something we could know reliably. Big bang theory was first theorized by a Roman Catholic priest and physicist, Georges Lemaître. To this day there are many believing Christians (and Jews and Muslims) in the sciences. Francis Collins is an evangelical Christian who is utterly convinced of evolution, led the Human Genome Project, and directed the National Institutes of Health.

How then did we get to equating science (and even "science") with secularism? Well, the so-called "conflict thesis" between science and faith became very influential in the United States in the nineteenth and twentieth centuries, thanks in large part to some unfair and poor scholarship from both the science side and the Christian side. Karl W. Giberson relates the frustrating history in his very readable book *Saving Darwin*.[24] (In fact, the pioneers of the conflict thesis, a chemist and a university founder and president, also bear a lot of responsibility for the widespread but easily refuted misconception that medieval Europeans believed the earth was flat. Europeans had known since the ancient Greeks that it was round, but thought it was smaller. Columbus sailed west to find a quick trade route to the Indies. Medieval and Renaissance books and stained-glass windows show round earths.) In many countries people don't see conflict, though the assumption that science and Christianity must conflict has spread through American scientists *and* missionaries to churches and scientific communities in other countries. Communists played a part too.[25]

How *do* science and faith relate? You'll hear different answers. Mine is that God's kingdom is the framework that puts all other things in their truest perspective. It is creation's goal and its broadest context. The kingdom is perceived through its "signs" (Acts 2:22–24), whereas the smaller domains of our many different forms of ordinary knowledge (scientific, historical, experiential, relational, and so on) are more obvious.[26] Knowledge of the kingdom—so theology—doesn't replace disciplines such as the sciences, and they don't replace it. No domain stands on its own; in fact, each of them needs the others, though not in the same ways.

Now if someone starts with the mindset that God is actively making every individual thing happen in the supposedly natural world—almost literally knitting new babies in their mothers' wombs—then it will seem "secularizing" to learn about natural processes that work on their own. If you've thought of every star, planet, and species as a special miracle from our Creator, then

24. You can find a brief history of the conflict thesis on Wikipedia.

25. See for instance Bukharin and Preobrazhensky, *The ABC of Communism*. DE: So too did fascists; I write about this in my forthcoming essay, "Why We Get Technology Wrong."

26. I have a full lecture on this on YouTube called "Framing and Painting."

astrophysics will make God seem less involved and further away. Newtonian mechanics spawned deism: the belief that God, like a watchmaker, built his creation, wound it up, and stood back. Then Darwinism and genetics relieved God of a lot of detailed watch-making work that Newtonian physics couldn't explain. Exclusive exposure to such learning may tempt someone to forget or deny God's obscure kingdom, despite its signs.

Scientific advance only leads to secularism if people jump to conclusions. Many have. Science's awesome predictive and explan-atory powers tempt us to neglect or explain away the kingdom's signs that lie beyond it, dismissing the wonders (Exhibit A: Jesus' resurrection) and other evidences that point beyond scientific domains and even conventional assumptions about history. The sciences' vast successes have blinded our culture to the reality that they don't and can't ascertain the *purposes* of the things they study, whereas God's purpose is the kingdom's focus. The impressive track record of science and technology lures us to glamorize the scientific method and think its knowledge is the best kind, where-as some of the most significant forms of knowledge lie beyond its horizon. Among these are interpersonal discovery—the relation-ships without which we would all die of loneliness—and history, which is what the gospel finally is: *news* about actual past events that bear upon the present and future. Scientific knowledge isn't better or worse than these. Each field apprehends its own domain and contributes to the whole of our knowledge.

When this is forgotten, bad things happen. Social Darwin-ism—evolutionary theory applied to politics, economics, and other social sciences—*did* fuel rugged individualism, exploitative capitalism, communism, fascism, and more. But those were what Douglas calls "science": misapplications of a biological paradigm (sometimes with evil intent) outside its proper domain, just as de-ism was a misapplication of Newtonian physics beyond its horizon. This happens everywhere. Physicists overreach when they take science's inability to infer purpose as evidence that the universe has no purpose. Sociologists overreach when the social character of knowledge and identity leads them to conclude that gender is entirely constructed. Economists and epidemiologists overreach when their models ignore human quirks and systems' complexities. Poets and playwrights overreach when they dismiss the idea that statistics could also yield insights into human nature. Political scientists overreach when they imagine laws and policies can bring about true justice. Theologians overreach when they spiritualize everything. Overreaching is a bad tree whose many bad fruits interfere with Christian faith. Ignoring genuine knowledge just makes Christian faith more isolated, distorted, and vulnerable, and its gospel less credible.

It's healthy to point out legitimate weaknesses and problems in all these disciplines. Every tradition grows and improves through criticism—including Christianity. Every church history book, and the New Testament writings themselves, show its intellectual tradition growing and developing all the way from Jesus' day to ours. For instance, all three of our chapters have cited Augustine because his insights were so helpful and influential, but he also developed some problematic concepts that churches have backed away from.

You've said a lot about natural forces and dynamics, but you obviously believe in miracles too. Do you affirm intelligent design? How was God involved in life's origins and history?

Evolutionary biology is ambitious. It looks for natural explanations of phenomena. For instance, why do human beings the world over have a visceral fear of spiders? Dogs go right up to spiders and sniff. Why aren't they afraid like we are? Biologists and anthropologists reason like this: Our ancestors could have evolved in an environment where that fear saved the lives of enough of the population that fearful survivors took over the gene pool. Logically, this would have happened before humans migrated all over the world, otherwise some populations would be arachnophobes and others wouldn't. Evolutionary psychologists reason the same way about human thinking and moral patterns.[27] I find much of their work helpful and drew on it earlier. Some label such explanations "evolutionary just-so stories," after Rudyard Kipling's *Just So Stories for Little Children*. Kipling spun fantastic tales of origins such as "How the Leopard Got Its Spots," but this label is no insult. Scientists offer hypotheses explaining what evidence we have, like the arachnophobia hypothesis above. Any reasonable possibility meets the challenge the phenomenon poses for evolutionary theory. So far, so good.

However, some people overreach, treating a conceivable hypothesis as compelling and a persuasive one as certain. Because it *could have* happened that way, therefore it *did*. Most scientists try to avoid this kind of overreaching, but not all. And overreachers and sensationalists are more likely to make a splash.[28] Sigmund Freud was a grand master. And while some popularizers do a great job of communicating complicated biological insights to general audiences—April certainly did that in her chapter—others dispense with caution and qualifications and treat hypotheses as facts.[29] The problem isn't the search for natural explanations! The problem is overconfidence in some circles (tribes?) in what is still a hypothesis—and may always be, since conclusive evidence of, say, arachnophobia's origins might be gone forever. Unfortunately, some skeptics of evolution collect oversold and obsolete evolutionary just-so stories, blow them out of proportion (which doesn't help their tribe's credibility), and treat them as evidence against

27. See, for example, Jonathan Haidt's insightful *The Righteous Mind*. I don't think Haidt overreaches.

28. Serra-Garcia and Gneezy say, "Papers that replicate are cited 153 times less, on average, than papers that do not." See "Nonreplicable Publications Are Cited More than Replicable Ones." Ironically, this study made headlines. Should we trust it?

29. Meanwhile, the public gets more "science" than science: a confusing mishmash of genuine scientific consensus, the purported consensus of zealous gatekeepers, honestly presented tentative findings, accurately stated and overstated provisional findings, and pseudoscientific quackery, all reported by PR flacks and journalists with often insufficient background knowledge, widely varying judgment, tribal instincts, and often axes to grind. We'll set the public information problem aside, but it dramatically affects perceptions and actions.

evolution as a whole, when generally they're just evidence of a field still working its way to deeper knowledge, or overconfidence at worst.[30]

Intelligent design faces the opposite temptation. Both ID and conventional biology focus on what's unexplained. But while conventional biology is committed to keep looking for any natural biological explanations, ID demotivates that search and trains people to *underreach*. Characterizing unexplained phenomena as circumstantial evidence of supernatural influence is a "science-stopper" that constructs a "God of the gaps." I jump back instinctively from a black widow and avoid a nasty bite, and I reflexively thank God for the reaction. This is well and good, but should I credit my arachnophobia to God's *particular* blessing of *special* protection to humanity? ID provides a ready framework—"design"—that discourages me from finding explanations that might weaken my appreciation for God and diminish his glory. After all, I might rationalize, God loves us, *and* God wants us to survive, multiply, fill, and subdue the earth. So shouldn't we directly credit God for arachnophobia? Isn't ID more faithful than evolution? I personally know believers, including science-minded engineers, who offer *anti*-evolutionary just-so stories for every feature of human nature. "Design" is right there to answer every "why" question. Habitually appealing to it would hobble me as a scientific thinker. Absence of evidence doesn't constitute evidence of absence. My answers will also make me look superstitious and foolish to people who don't think this way, weaken my testimony's credibility, and expose my own faith to quite reasonable challenges. *Reaching*—neither overreaching nor underreaching, but reaching responsibly in all the ways we're equipped—best helps us understand our world and its origins.

So how *was* God involved in the origins and development of creation? Biblical faith praises God for special acts conjuring

30. AC: It's incredibly challenging for me to remain gracious when someone uses an obsolete, overblown evolutionary explanation to reject evolution altogether. And when I'm told that "there is no evidence for evolution," I try very hard not to take this as an insult to my intellect and my chosen field of biology.

things out of nothing (Hebrews 11:3) and making new things from old (John 2:1–11), as well for ongoing natural phenomena such as the seasons, water and life cycles, and organic fertility. How much of this was natural, and how much supernatural? Here I look to scientists of faith for evidence, and they find few true unexplainables. Among those are the big bang of course, as well as our "fine-tuned universe" (the so-called anthropic principle) and perhaps the rise of altruism (love of strangers and even enemies at the cost of one's own life). And of course it's scientists' job to keep prodding these, looking for natural explanations and insights.

I respect intelligent design's openness to supernatural explanations for things and, at its best anyway, its critiques of past scientific overreach and "scientific" distortion. Sometimes things happen that won't have natural explanations. There is no theological reason to rule out special interventions from God along life's way. But is something supernatural just because we can't explain it?[31] We don't know until we try, and over the years we've found a lot of natural explanations for things once considered supernatural. ID fights overconfidence with underconfidence, and dissuades people from trying. I'll put it this way: I'm happy if my family's obstetrician agrees with me that our children are gifts from God, but I don't want an obstetrician who thinks God does the pushing.

Doesn't putting our trust in claims that seem at odds with biblical faith fail God's test of faith?

This question reflects the conflict thesis's power in our imaginations, as well our culture's legacy of overreaching and under-reaching. It's not unfaithful to assume a natural explanation for something. Believers and nonbelievers alike have always expected life to follow natural patterns. It ought to take something special to convince us of a miracle. Jesus didn't call Peter out from the boat

---

31. AC: I would say "can't explain it *yet*." While scientists may not be able to explain all of creation, we continue to try, and this does not diminish our faith or dishonor God.

until Peter had seen him walking on water! He gave Peter *reason* to trust.

When people deny Jesus' virginal conception or resurrection because *those kinds of things don't happen*, they're overreaching in the direction of naturalism. But to persist in thinking something's a miracle when the evidence points the other way is also overreaching, in the direction of spiritualizing. Jesus usually used a boat, and not because he lacked faith. Nature's ordinary course is God's doing, and it's good.

How often did Jesus say "behold"? Faith isn't closing our eyes and leaping blind into contradiction. The Lord didn't tell people to do that. He *opened* eyes and taught his people to see. Faith is trust that springs from God's demonstrated trustworthiness. Many Psalms look back to the exodus's miracles *and* creation's goodness as proof God is trustworthy. They don't see a conflict. Likewise, the church looks to Christ's resurrection *and* God's ordinary providence. We have the Spirit's gift of discernment to nurture our judgment in interpreting everything from scientific hypotheses and findings to the Scriptures and current events.

Am I supposed to be encouraged by your answers?! They make the gospel and Bible so much more complicated and elitist. Aren't those for everyone? Earlier generations couldn't know all this information, and many today still aren't aware of it.

It's true that the Christian church hasn't needed geology, astrophysics, or biology to know and love the God of Israel. None of this knowledge was or is God's priority. But now that it's here, we have to—we *get* to—deal with it.

The Bible's writings *are* for all Israel and all churches to hear and understand. But it has always taken people with learning to relate them to common people, especially back when literacy was low. So its books are written for general audiences *through* elites. They're not *elitist*. Protestants so valued the power of Scripture that they labored to make "elite" literacy universal. Widespread literacy spreads the responsibility of knowing how to read the Bible well.

This takes effort even beyond adjusting for historical, cultural, and scientific gaps. Hebrew narrative is coy and subtle. Apocalyptic writing is purposefully obscure and demands familiarity with its devices and symbols. Luke is written for a more learned audience than Mark; John's work is highly structured and symbolic; and Paul's reasoning in Romans is much more elaborate than in 1 Corinthians. It's our privilege to reckon with Scripture's sophistication and mine its richness: "Every scribe who has been trained for the kingdom of heaven is like a householder who brings out of his treasure what is new and what is old" (Matthew 13:52).

For all this complexity, the Bible's obvious messages do stand out clearly. I serve as a volunteer chaplain at our local county jail. Some of the men I teach and worship with are barely literate, but their grasp of the gospel is firm and insightful. They appreciate it when I can use my education to clear up confusion about some details, but I learn from them as much as they learn from me. We differ on all sorts of topics including evolution, but we all share the same faith, and the Lord is with all of us until the end of the age.[32]

So I *do* mean to encourage you! I don't expect you to agree with all I've said here, but I hope I've helped you see the Christian faith as *both* robust and sure of its life-giving core message and Scriptures, *and* hospitable to the scientific tradition that has brought limited but awe-inspiring and massively fruitful insights into God's creation. Behold and explore them both with hope and joy!

For further reading:

Timoth Keller, "Creation, Evolution, and Christian Laypeople."
N. T. Wright, "If Creation is Through Christ, Evolution is What You Would Expect."
Karl Giberson, *Saving Darwin.*

32. DE: Right. Even if we differ on evolution, or on details about Adam and Eve, as long as we hold fast to the faith God gives us, we will be rescued from this broken world.

Bibliography

Almécija, Sergio, et al. "Fossil Apes and Human Evolution." *Science* 372:6542 (May 7, 2021). science.sciencemag.org/content/372/6542/eabb4363.

Applegate, Kathryn. "Understanding Randomness." BioLogos, April 8, 2010. biologos.org/articles/understanding-randomness.

———. "Why I Think Adam was a Real Person in History." *BioLogos*, June 11, 2018. biologos.org/articles/why-i-think-adam-was-a-real-person-in-history.

Asher, Robert. *Evolution and Belief: Confessions of a Religious Paleontologist.* Cambridge: Cambridge University Press, 2012.

Augustine. *Exposition on Psalm 139.* www.newadvent.org/fathers/1801139.htm.

———. *The Literal Meaning of Genesis: Volume I, Books 1–6.* Translated by John Hammond Taylor. Mahwah, NJ: Paulist, 1982.

———. *Teaching Christianity: De Doctrina Christiana.* Translated by Edmund Hill. New York: New City, 1995.

BioLogos.org. "Were Adam and Eve Historical Figures?" biologos.org/common-questions/were-adam-and-eve-historical-figures.

———. "Famous Christians Who Believed Evolution is Compatible with Christian Faith." biologos.org/articles/famous-christians-who-believed-evolution-is-compatible-with-christian-faith.

———. "B. B. Warfield, Biblical Inerrancy, and Evolution." biologos.org/articles/b-b-warfield-biblical-inerrancy-and-evolution.

Bonnette, Dennis. "The Rational Credibility of a Literal Adam and Eve." *Espíritu* 64 (2015) 303–20.

Brainard, Jeffrey, and Jia You. "What a Massive Database of Retracted Papers Reveals about Science Publishing's 'Death Penalty.'" *Science*, October 25, 2018. www.sciencemag.org/news/2018/10/what-massive-database-retracted-papers-reveals-about-science-publishing-s-death-penalty.

Branch, Glenn. "What's Wrong with 'Belief in Evolution?'" ncse.ngo/whats-wrong-belief-evolution-part-1.

Bukharin, N. I., and E. Preobrazhensky. *The ABC of Communism.* New York: Penguin, 1969.

Calvin, John. *Commentaries on the Epistles to Timothy, Titus, and Philemon.* Translated by William Pringle. Edinburgh: Calvin Translation Society, 1856.

Davidson, Gregg. *Friend of Science, Friend of Faith: Listening to God in His Works and Word.* Grand Rapids: Kregel, 2019.

Enns, Peter. "Genesis 1 and a Babylonian Creation Story." biologos.org/articles/genesis-1-and-a-babylonian-creation-story.

Estes, Douglas. *Braving the Future: Christian Faith in a World of Limitless Tech.* Harrisonburg, PA: Herald, 2018.

———. "Why We Get Technology Wrong." Forthcoming in *Technē: Christian Visions of Technology.* Edited by Gerald Hiestand and Todd Wilson. Eugene, OR: Cascade, 2022.

Estes, Douglas, ed. *The Tree of Life.* Themes in Biblical Narrative 27. Leiden: Brill, 2020.

Falk, Darrel. "Chromosome 2 Part 1: Evidence for an Evolutionary Creation." www.youtube.com/watch?v=4oQbU4O8--E.

———. *Coming to Peace with Science.* Downers Grove, IL: InterVarsity, 2004.

Giberson, Karl W. *Saving Darwin: How to Be a Christian and Believe in Evolution.* New York: HarperCollins, 2008.

Gilad, Yoav, Orna Man, Svante Pääbo, and Doron Lancet. "Human Specific Loss of Olfactory Receptor Genes." *Proceedings of the National Academy of Sciences* 100(6) (2003) 3324–27.

Gingerich, Owen. *God's Universe.* Cambridge, MA: Belknap Press of Harvard University Press, 2006.

Haarsma, Deborah B., and Loren D. Haarsma. *Origins: Christian Perspectives on Creation, Evolution, and Intelligent Design.* Rev. ed. Grand Rapids: Faith Alive, 2011.

Haidt, Jonathan. *The Righteous Mind: Why Good People Are Divided by Politics and Religion.* New York: Vintage, 2013.

Harari, Yuval Noah. *Homo Deus: A Brief History of Tomorrow.* New York: Harper, 2017.

Harvard Medical School. "The Evolution of Bacteria on a 'Mega-Plate' Petri Dish (Kishony Lab). www.youtube.com/watch?v=plVk4NVIUh8.

Kaiser, Walter C., and Moisés Silva. *An Introduction to Biblical Hermeneutics: The Search for Meaning.* Grand Rapids: Zondervan, 1994.

Keller, Timothy. "Creation, Evolution, and Christian Laypeople." BioLogos, 2012. biologos.org/articles/creation-evolution-and-christian-laypeople.

———. *The Reason for God: Belief in an Age of Skepticism.* New York: Penguin, 2018.

LeFebvre, Michael. *The Liturgy of Creation: Understanding Calendars in Old Testament Context.* Downers Grove, IL: InterVarsity, 2019.

Lodahl, Michael, and April Cordero Maskiewicz. *Renewal in Love: Living Holy Lives in God's Good Creation.* Kansas City, MO: Beacon Hill, 2014.

Longman, Tremper, III, and John H. Walton. *The Lost World of the Flood.* Downers Grove, IL: InterVarsity, 2018.

Murphy, Nancey. "Science and Society." In *Witness: Systematic Theology Volume 3*, by James Wm. McClendon Jr., 99–131. Nashville: Abingdon, 2010.

National Academy of Sciences. "Frequently Asked Questions." www.nationalacademies.org/evolution/faq.

National Center for Education Statistics. *Digest of Education Statistics (2019)*. nces.ed.gov/programs/digest/d19/tables/dt19_322.10.asp.

Nichols, Amanda J., and Myron A. Penner. "Nuclear Chemistry and Medicine: Why 'Young-Earthers' Cannot Have It Both Ways." *Perspectives on Science and Christian Faith* 71:4 (December 2019) 203–17. www.asa3.org/ASA/PSCF/2019/PSCF12-19Complete.pdf

Peels, Rik. "A Conceptual Map of Scientism." In *Scientism: Prospects and Problems*, edited by Jeroen de Ridder, Rik Peels, and René van Woudenberg, 28–56. Oxford: Oxford University Press, 2018.

Pew Research Center. "Major Gaps Between the Public, Scientists on Key Issues." July 1, 2015. www.pewresearch.org/internet/interactives/public-scientists-opinion-gap.

———. "Scientific Achievements Less Prominent Than a Decade Ago: Public Praises Science; Scientists Fault Public, Media." July 9, 2009. www.pewresearch.org/wp-content/uploads/sites/4/legacy-pdf/528.pdf.

———. "Strong Role of Religion in Views About Evolution and Perceptions of Scientific Consensus." October 22, 2015. www.pewresearch.org/science/2015/10/22/strong-role-of-religion-in-views-about-evolution-and-perceptions-of-scientific-consensus/.

Principe, Lawrence M. "Scientism and the Religion of Science." In *Scientism: The New Orthodoxy*, edited by Richard N. Williams and Daniel N. Robinson, 41–62. London: Bloomsbury, 2015.

Rau, Gerald. *Mapping the Origins Debate: Six Models of the Beginning of Everything*. Downers Grove, IL: InterVarsity, 2012.

Rosenberg, Joel. "Biblical Narrative." In *Back to the Sources: Reading the Classic Jewish Texts*, edited by Barry W. Holtz, 31–82. New York: Simon & Schuster, 1984.

Sacks, Jonathan. "Rabbi Sacks on 'The Great Partnership.'" rabbisacks.org/great-partnership.

Scientific American. "Evolution." www.scientificamerican.com/evolution.

Serra-Garcia, Marta, and Uri Gneezy. "Nonreplicable Publications Are Cited More than Replicable Ones." *Science Advances* 7:21 (May 21, 2021). advances.sciencemag.org/content/7/21/eabd1705.

Shubin, Neil. *Your Inner Fish*. New York: Vintage, 2009.

Swamidass, S. Joshua. *The Genealogical Adam and Eve: The Surprising Science of Universal Ancestry*. Downers Grove, IL: InterVarsity, 2019.

U.S. Census Bureau. "Educational Attainment in the United States: 2019." www.census.gov/newsroom/press-releases/2020/educational-attainment.html.

Venema, Dennis. "How Language Evolution Helps Us Understand Human Evolution." BioLogos, May 22, 2014. biologos.org/articles/series/

evolution-basics/how-language-evolution-helps-us-understand-human-evolution.

Wei-Haas, Maya. "Multiple Lines of Mysterious Ancient Humans Interbred with Us." *National Geographic,* April 11, 2019. www.nationalgeographic.com/science/article/enigmatic-human-relative-outlived-neanderthals.

Wikipedia.org. "Conflict Thesis." en.wikipedia.org/wiki/Conflict_thesis.

Work, Telford. "Framing and Painting: Relating Domains of Knowledge." youtube/ZOSLOtiUhZ4.

Wright, N. T. "If Creation is Through Christ, Evolution is What You Would Expect." BioLogos, 2017. biologos.org/resources/if-creation-is-through-christ-evolution-is-what-you-would-expect.